Instructor's Guide

Trigonometry

SECOND EDITION

and

Trigonometry

Hybrid

James Stewart
McMaster University and University of Toronto

Lothar Redlin
The Pennsylvania State University

Saleem Watson
California State University, Long Beach

Prepared by

Douglas Shaw
University of Northern Iowa

BROOKS/COLE
CENGAGE Learning

Australia • Brazil • Japan • Korea • Mexico • Singapore • Spain • United Kingdom • United States

BROOKS/COLE
CENGAGE Learning

For product information and technology assistance, contact us at **Cengage Learning Customer & Sales Support, 1-800-354-9706**

For permission to use material from this text or product, submit all requests online at **www.cengage.com/permissions**
Further permissions questions can be emailed to **permissionrequest@cengage.com**

ISBN-13: 978-1-111-57476-5
ISBN-10: 1-111-57476-6

Brooks/Cole
20 Davis Drive
Belmont, CA 94002-3098
USA

Cengage Learning is a leading provider of customized learning solutions with office locations around the globe, including Singapore, the United Kingdom, Australia, Mexico, Brazil, and Japan. Locate your local office at: **www.cengage.com/global**

Cengage Learning products are represented in Canada by Nelson Education, Ltd.

To learn more about Brooks/Cole, visit **www.cengage.com/brookscole**

Purchase any of our products at your local college store or at our preferred online store **www.cengagebrain.com**

Printed in Milton Keynes UK
1 2 3 4 5 30 29 28 27 26

CONTENTS

PREFACE

The purpose of this *Instructor's Guide* is to save you time while helping you to teach a nice, honest, interesting, student-centered course. For each section, there are suggested additions to your lecture that can supplement (but not replace) things like graphing $\sin(3x + \pi)$ or simplifying $2\sin x \cos x$. Lecturing is not your only option, of course. This guide provides group activities, ready for reproducing, that will allow your students to discover and explore the concepts of trigonometry. You may find that your classes become more "fun", but I assure you that this unfortunate by-product of an engaged student population can't always be avoided.

This guide was designed to be used with *Trigonometry, Second Edition* as a source of both supplementary and complementary material. Depending on your preference, you can either occasionally glance through the *Guide* for content ideas and alternate approaches, or you can use it as a major component in planning your day-to-day classes. In addition to lecture materials and group activities, each section has examples, sample homework assignments, and reading quizzes.

For many students who do go on to take calculus, one major stumbling block for many is trigonometry. Therefore, teaching this subject is an important and noble task. It is my hope that this book will make that task a little easier.

This guide could not have been completed without the help of many people. For giving me this opportunity, I thank Jim Stewart, Lothar Redlin, Saleem Watson, Bob Pirtle, and Gary Whalen. Cynthia Ashton has been a kind and patient editor, and I'm thankful for her work on this project. Thanks to Job Evers, Ken Doss, Kate Degner, and especially Melissa Pfohl for their proofreading and suggestions for previous editions of this guide. Kate Degner gave this edition a very thorough read, and many excellent changes came as a result of her ideas. My wife Laurel's support and love throughout this and all of my projects have been key to their completion. In addition, she did the research for the exponent group work, and proofread many others. If you like the way this book looks, if you admire the clarity of the graphics and the smoothness of its design, join me in thanking Andy Bulman-Fleming, one of the best typesetters in the business.

This book is dedicated to mathematics teacher Slade Hovick, whose height is dwarfed by the greatness of his heart.

Douglas Shaw
University of Northern Iowa

HOW TO USE THE INSTRUCTOR'S GUIDE

For each section of *Trigonometry, Second Edition*, this *Instructor's Guide* provides information on the items listed below.

1. **Suggested Time and Emphasis** These suggestions assume that the class is fifty minutes long. They also advise whether or not the material is essential to the rest of the course. If a section is labeled "optional", the time range given is the amount of time for the material in the event that it is covered.
2. **Points to Stress** This is a short summary of the big ideas to be covered.
3. **Sample Questions** Some instructors have reported that they like to open or close class by handing out a single question, either as a quiz or to start a discussion. Two types are included:
   - **Text Question:** This question is designed for students who have done the reading, but haven't yet seen the material in class. These questions can be used to help ensure that the students are reading the textbook carefully.
   - **Drill Question:** These questions are designed to be straightforward "right down the middle" questions for students who have tried, but not necessarily mastered, the material.
4. **In-Class Materials** These suggestions are meant to work along with the text to create a classroom atmosphere of experimentation and inquiry.
5. **Examples** These are routine examples with all the computations worked out, designed to save a bit of time in class preparation.
6. **Group Work** One of the main difficulties instructors have in presenting group work to their classes is that of choosing an appropriate group task. Suggestions for implementation and answers to the group activities are provided first, followed by photocopy-ready handouts on separate pages. This guide's main philosophy of group work is that there should be a solid introduction to each exercise ("What are we supposed to do?") and good closure before class is dismissed ("Why did we just do that?")
7. **Homework Problems** For each section, a set of essential **Core Exercises** (a bare minimum set of homework problems) is provided. Using this core set as a base, a **Sample Assignment** is suggested.

TIPS ON IN-CLASS GROUP WORK

This *Instructor's Guide* gives classroom-tested group work activities for every section of *Trigonometry, Second Edition*. One reason for the popularity of in-class group work is that *it is effective*. When students are engaged in doing mathematics, and talking about mathematics with others, they tend to learn better and retain the material longer. Think back to your own career: didn't you learn a lot of mathematics when you began teaching it to other people? Many skeptics experiment by trying group work for one semester, and then they get hooked. Pick a group activity from the guide that you like, make some photocopies, and dive in!

1. Mechanics

Books and seminars on in-class group work abound. I have conducted many such seminars myself. What follows are some tips to give you a good start:

(a) **Do it on the first day.**

The sources all agree on this one. If you want your students to believe that group work is an important part of the course, you have to start them on the first day. My rule of thumb is "at least three times the first week, and then at least once a week thereafter." I mention this first because it is the most important.

(b) **Make them move.**

Ideally, students should be eye-to-eye and knee-to-knee. If this isn't possible, do the best you can. But it is important to have them move. If your groups are randomly selected, then they will have to get up and sit in a different chair. If your groups are organized by where they are seated in the classroom, make them move their chairs so they face each other. There needs to be a "break" between sitting-and-writing mode and talking-to-colleagues mode.

(c) **Use the ideal group size.**

Research has shown that the ideal group size is three students, with four-student groups next. I like to use groups of four; if one of them is absent (physically or otherwise), the group still has three participating members.

(d) **Fixed versus random groups.**

There is a lot of disagreement here. Fixed groups allow each student to find her or his niche, and allow you to be thoughtful when you assign groups or reassign them after exams. Random groups allow students to have the experience of working with a variety of people. I believe the best thing to do is to try both methods, and see which works best for you and your students.

(e) **Should students hand in their work?**

The advantage of handing in group works is accountability. My philosophy is that I want the group work to have obvious, intrinsic benefit. I try to make the experience such that it is obvious to the student that they get a lot out of participating, so I don't need the threat of "I'm grading this" to get them to focus. I sometimes have the students hand in the group work, but only as a last resort.

2. Closure

As stated above, I want my students to understand the value of working together actively in their groups. Once you win this battle, you will find that a lot of motivation and discipline problems simply go away. I've found the best way to ensure that the students understand why they've done an activity is to tell them. The students should leave the room having seen the solutions and knowing why they did that particular activity. You can have the students present answers or present them yourself, whatever suits your teaching style. I've had success with having groups write their results on transparencies and present them to the class (after I've checked their accuracy).

Here is another way to think about closure: Once in a while, give a future homework problem out as a group work. When the students realize that participating fully in the group work helps them in the homework, they get a very solid feeling about the whole process.

3. Introduction

The most important part of a group activity, in my opinion, is closure. The second most important is the introduction. A big killer of group work is that awful time between you telling your students they can start, and the first move of pencil on paper—the "what on earth do we do now?" moment. A good introduction should be focused on getting them past that moment. You don't want to give too much away, but you also don't want to throw them into the deep end of the swimming pool. In some classes, you will have to say very little, and in some you may have to do the first problem with them. Experiment with your introductions, but never neglect them.

4. Help when you are needed

Some group work methods involve giving absolutely no help when the students are working. Again, you will have to find what is best for you. If you give help too freely, the students have no incentive to talk to each other. If you are too stingy, the students can wind up frustrated. When a student asks me for help, I first ask the group what they think, and if it is clear they are all stuck at the same point, I give a hint.

5. Make understanding a goal in itself

Convey to the students (again, directness is a virtue here) that their goal is not just to get the answer written down, but to ensure that every student in their group understands the answer. Their work is not done until they are sure that every one of their colleagues can leave the room knowing how to do the problem. You don't have to sell every single student on this idea for it to work.

6. Bring it back when you can

Many of the group works in this guide foreshadow future material. When you are lecturing, try to make reference to past group works when it is appropriate. You will find that your students more easily recall a particular problem they discussed with their friends than a particular statement that you made during a lecture.

The above is just the tip of the iceberg. There are plenty of resources available, both online and in print. Don't be intimidated by the literature—start it on the first day of the next semester, and once you are into it, you may naturally want to read what other people have to say!

HOW TO IMPLEMENT THE DISCOVERY PROJECTS

One exciting yet intimidating aspect of teaching a course is projects. An extended assignment gives students the chance to take a focused problem or project and explore it in depth—making conjectures, discussing them, eventually drawing conclusions and writing them up in a clear, precise format. *Trigonometry, Second Edition* has many Discovery Projects throughout. They are excellent and well thought out, and should be explored if possible. Here are some tips on ensuring that your students have a successful experience.

Time Students should have two to three weeks to work on any extended out-of-class assignment. This is not because they will need all this time to complete it! But a fifteen-to-twenty-day deadline allows the students to be flexible in structuring their time wisely, and allows the instructors to apply fairly strict standards in grading the work.

Groups Students usually work in teams and are expected to have team meetings. The main problem students have in setting up these meetings is scheduling. Four randomly selected students will probably find it very hard to get together for more than a few hours, which may not be sufficient. One way to help your students is to clearly specify a minimum number of meetings, and have one or all group members turn in summaries of what was accomplished at each meeting. A good first grouping may be by location.

Studies have shown that the optimal group size is three people, followed by four, then two. I advocate groups of four whenever possible. That way, if someone doesn't show up to a team meeting, there are still three people there to discuss the problems.

Before the first project, students should discuss the different roles that are assumed in a team. Who will be responsible for keeping people informed of where and when they meet? Who will be responsible for making sure that the final copy of the report is all together when it is supposed to be? These types of jobs can be assigned within the team, or by the teacher at the outset.

Tell the students that you will be grading on both content and presentation. They should gear their work toward an audience that is bright, but not necessarily up-to-speed on this problem. For example, they can think of themselves as professional mathematicians writing for a manager, or as research assistants writing for a professor who is not necessarily a mathematician.

If the students are expected to put some effort into the project, it is important to let them know that some effort was put into the grading. Both form and content should be commented on, and recognition of good aspects of their work should be included along with criticism.

One way to help ensure cooperation is to let the students know that there will be an exam question based on the project. If every member of the group does well on that particular question, then they can all get a bonus, either on the exam or on the project grade.

Providing Assistance Make sure that the students know when you are available to help them, and what kind of help you are willing to provide. Students may be required to hand in a rough draft ten days before the due date, to give them a little more structure and to make sure they have a solid week to write up the assignment.

Individual Accountability It is important that the students are individually accountable for the output of their group. Giving each student a different grade is a dangerous solution, because it does not necessarily encourage the students to discuss the material, and may actually discourage their working together. A better alternative might be to create a feedback form. If the students are given a copy of the feedback form ahead of time, and they know that their future group placement will be based on what they do in their present group, then they are given an incentive to work hard. One surprising result is that when a group consists of students who were previously slackers, that group often does quite well. The exam question idea discussed earlier also gives individuals an incentive to keep up with their colleagues.

HOW TO USE THE REVIEW SECTIONS AND CHAPTER TESTS

Review sections for chapters of a textbook are often assigned to students the weekend before a test, but never graded. Students realize they won't be evaluated on this work and often skip the exercises in place of studying previous quizzes and glancing at old homework. A more useful activity for students is to use the review sections in *Trigonometry, Second Edition* to discover their precise areas of difficulty. Implemented carefully, these are a useful resource for the students, particularly for helping them to retain the skills and concepts they've learned. To encourage more student usage, try the following alternatives:

1. As the course goes on, note the types of exercises which caused difficulties. During a review session, assign students to work on similar exercises in the review sections and go over them at the end of class. Also assign exercises reminiscent of the ones you plan to have on the exam.
2. Use the review section problems to create a game. For instance, break students into groups and have a contest where the group that correctly answers the most randomly picked review questions "wins". One fun technique is to create a math "bingo" game. Give each group a 5×5 grid with answers to review problems. If you laminate the cards, and give the students dry-erase markers, then you can use them year after year. Randomly pick review problems, and write the questions on the board. Make sure that for a group to win, they must not only have the correct answers to the problems, but be able to give sound explanations as to how they got the answers.
3. A very "low-maintenance" way to give students an incentive to look at the chapter tests is to use one of the problems, verbatim, for an exam question, and make no secret of your intention to do so. It is important that students have an opportunity to get answers to any questions they have on the chapter tests before the exam is given; otherwise, this technique loses a great deal of its value.

1 FUNCTIONS AND GRAPHS

1.1 COORDINATE GEOMETRY

▼ Suggested Time and Emphasis

1– 2 classes. Essential material.

▼ Points to Stress

1. The distance and midpoint formulas.
2. The relationship between a two-variable equation and its graph. (This is the most crucial concept.)
3. Sketching graphs by plotting points, using intercepts and symmetry.
4. Equations of circles and their graphs, including the technique of completing the square.

▼ Sample Questions

- **Text Question:** Circle the type of types of symmetry exhibited by each of the following figures.

(a)

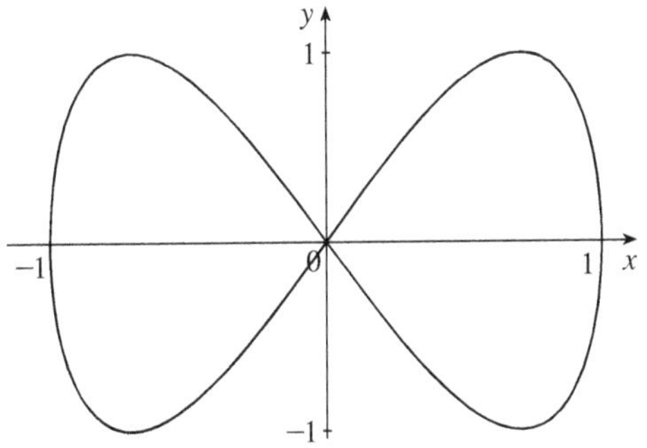

(i) About the x-axis
(ii) About the y-axis
(iii) About the origin

(b)

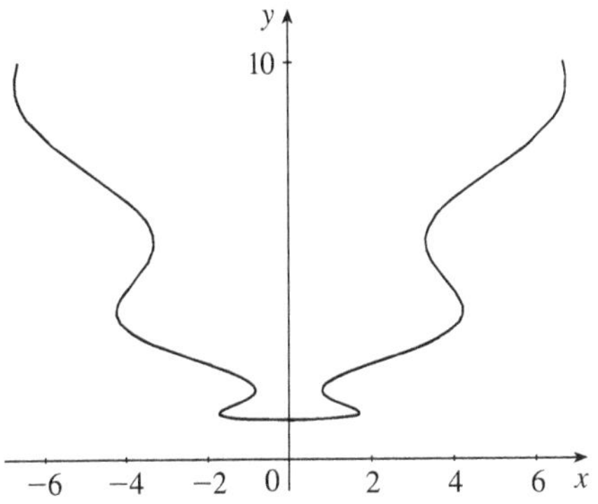

(i) About the x-axis
(ii) About the y-axis
(iii) About the origin

Answer: (a) (i), (ii), and (iii) **(b)** (ii)

- **Drill Question:** Graph the circle with equation $x^2 + 2x + y^2 - 4y = 20$.

Answer:

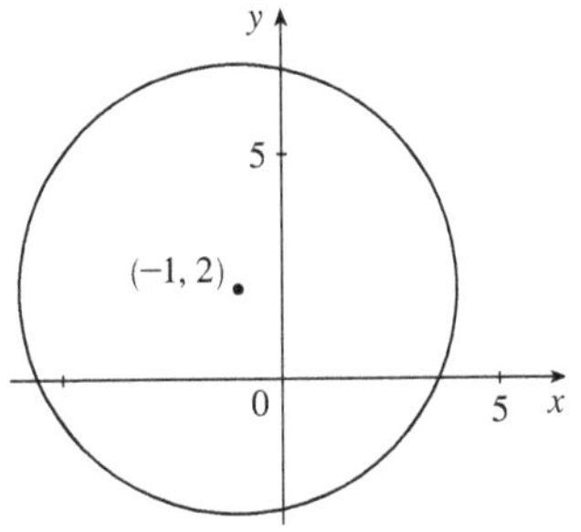

▼ In-Class Materials

- Obtain a transparency of a local map, and overlay it with an appropriate grid. Try to estimate the walking distance between the classroom and various points of interest. Then use the distance formula to find the distance "as the crow flies". This is discussed in Exercise 113 in the text.

- Assume that a meter stick is situated as in the diagram at right. Ask the students to find the 50 cm mark, and then the 25 cm mark.

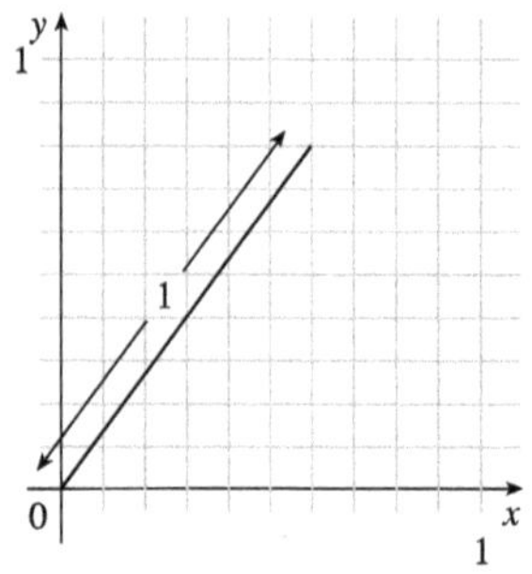

- I have always found this interesting, and your students may, too. Many modern cities are laid out in a grid similar to René Descartes' coordinate system. Paris, however, looks something like this:

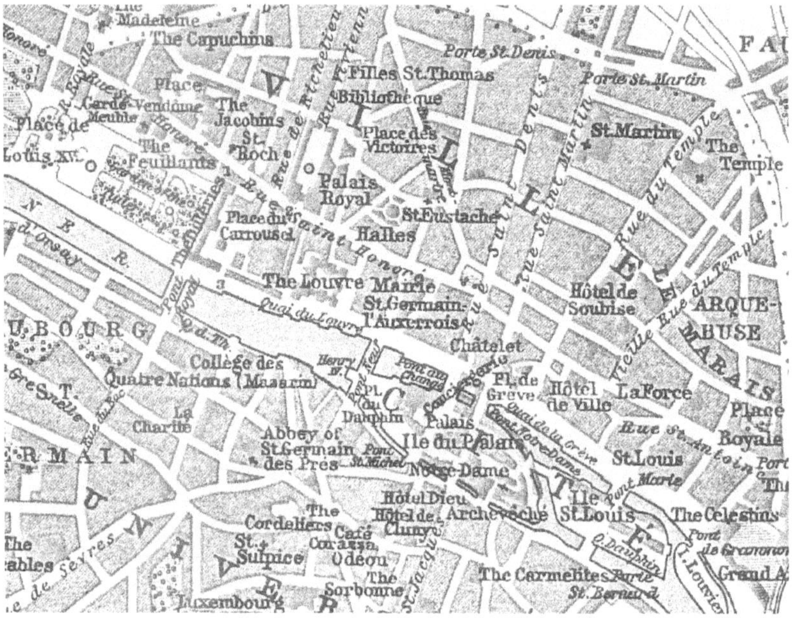

One can imagine René Descartes looking at such a map and saying, "There *must* to be a more organized way to arrange points in a plane!"

- Graph $y = ||x| - 1| - 2$ by plotting points. This is a nice curve because it is easy to get points, but students will not immediately see what the curve should look like. Have different students obtain different points, and plot them on a common graph on the board.

Answer:

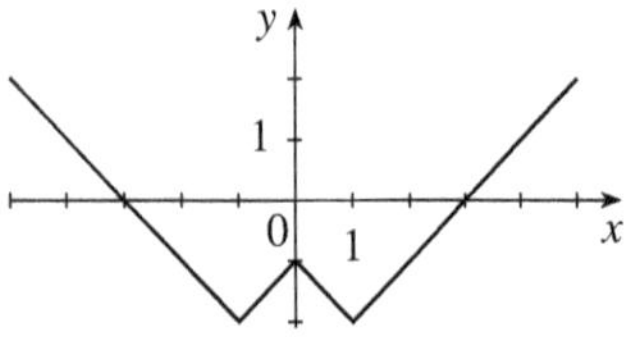

- Emphasize that the equation $(x - 2)^2 + (y + 3)^2 = 100$ and the equation $x^2 - 4x + y^2 + 6y = -3$ have the same graph — the same points satisfy both equations. Perhaps test $(2, 7)$ and $(1, 1)$ in both equations and notice that $(2, 7)$ satisfies both, and $(1, 1)$ fails to satisfy either. The only difference between the two equations is the form: it is easier to work with the first than the second.

▼ Examples

- Investigate the trapezoidal region given by the inequalities $x > 2$, $y > \frac{3}{2}x - 3$, $y > -x + 7$, and $y < -x + 12$.
- A curve with a lot of intercepts: $y = x^5 - 15x^4 + 85x^3 - 225x^2 + 274x - 120$ looks frightening until students are told it factors to $y = (x-1)(x-2)(x-3)(x-4)(x-5)$.

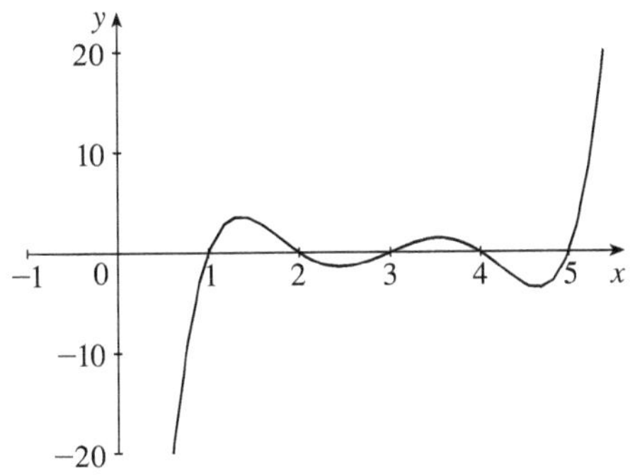

▼ Group Work 1: Taxicab Geometry

Taxicab geometry was first invented in the late 1800s, and is an accessible example of non-Euclidean geometry. In taxicab geometry, we assume a grid where we are allowed to travel only along the roads, and we define "distance" as driving distance. For example, in taxicab geometry, the distance between the point $(0,0)$ and the point $(5,6)$ is 11, because we would have to drive five blocks north and six blocks east to get from $(0,0)$ to $(5,6)$. There are analogues to circles, parabolas, and all the conic sections in taxicab geometry. This activity explores the distance formula in this context, and then checks to see whether the student really understands the definition of a circle.

If Chapter 11is to be covered, note at that time that one can draw taxicab ellipses, hyperbolas, and parabolas.

Answers:

1. No, you are a four-mile drive away.

2.

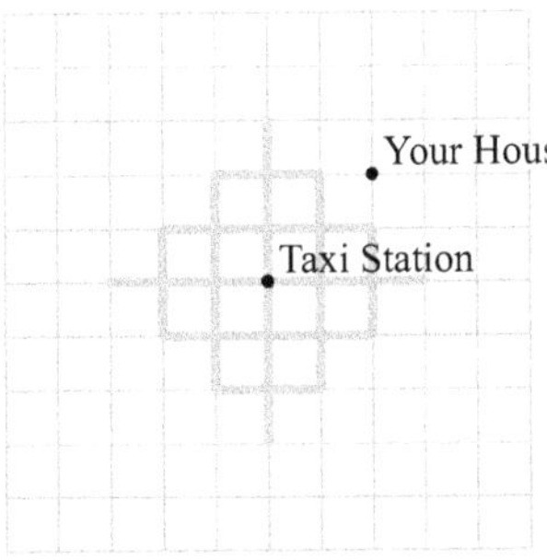

1 unit = 1 mile

3.

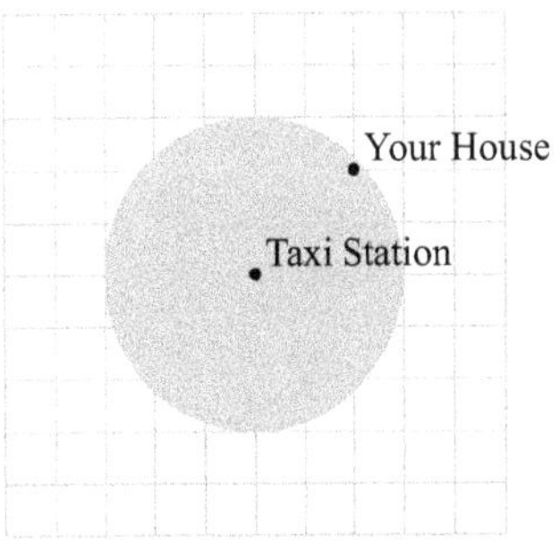

1 unit = 1 mile

4. $d = |x_1 - x_2| + |y_1 - y_2|$

▼ Group Work 2: Discovering the Shift

This group work foreshadows Section 2.5: Transformations of Functions. The curves are obtainable by plotting points, although other techniques may be used. The first part deals with functions, although it doesn't use the term "function". The second part deals with $y^2 = x$, whose graph is also easily obtainable by plotting points.

Answers (Part One)

1. $y = x^2$: x- and y-intercepts $(0, 0)$
   $y = (x-1)^2$: x-intercept $(1, 0)$, y-intercept $(0, 1)$
   $y = (x-2)^2$: x-intercept $(2, 0)$, y-intercept $(0, 4)$
   $y = (x-3)^2$: x-intercept $(3, 0)$, y-intercept $(0, 9)$
   $y = (x-4)^2$: x-intercept $(4, 0)$, y-intercept $(0, 16)$

2.
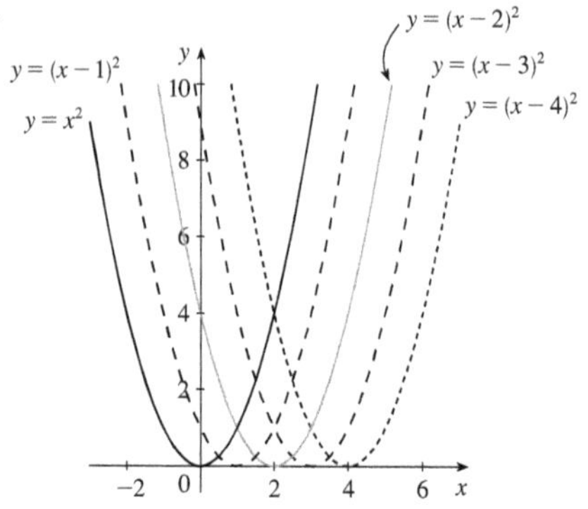

3. $y = \frac{1}{2}x$: x- and y-intercepts $(0, 0)$
   $y = \frac{1}{2}(x-1)$: x-intercept $(1, 0)$, y-intercept $\left(0, -\frac{1}{2}\right)$
   $y = \frac{1}{2}(x-2)$: x-intercept $(2, 0)$, y-intercept $(0, -1)$
   $y = \frac{1}{2}(x-3)$: x-intercept $(3, 0)$, y-intercept $\left(0, -\frac{3}{2}\right)$

4.
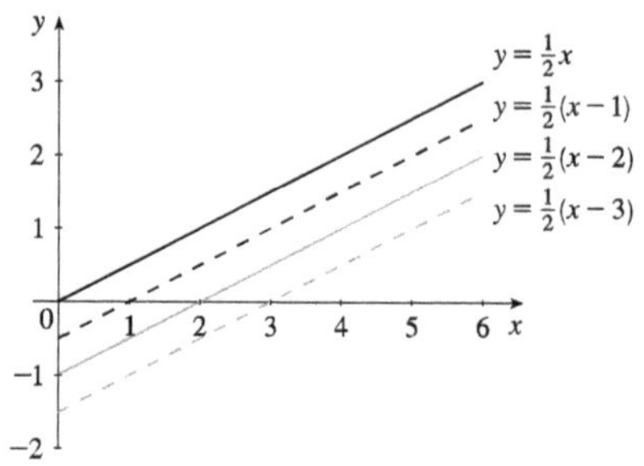

5.
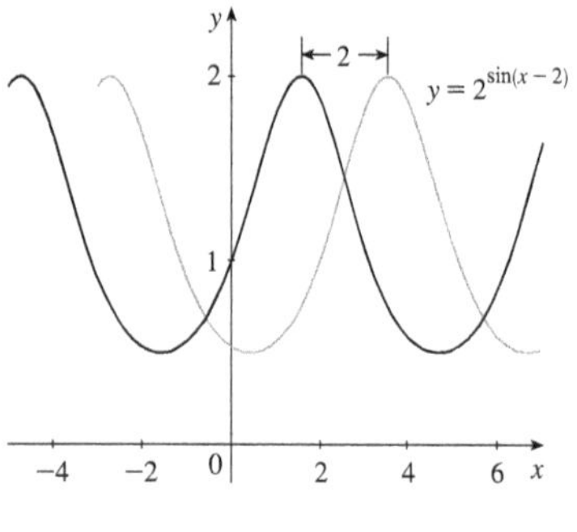

6. Replacing x with $x - k$ shifts a graph k units to the right.

Answers (Part Two)

1. $y^2 = x$: x- and y-intercepts $(0, 0)$.
   $(y-1)^2 = x$: x-intercept $(0, 1)$, y-intercept $(1, 0)$
   $(y-2)^2 = x$: x-intercept $(0, 2)$, y-intercept $(4, 0)$
   $(y-3)^2 = x$: x-intercept $(0, 3)$, y-intercept $(9, 0)$
   $(y-4)^2 = x$: x-intercept $(0, 4)$, y-intercept $(16, 0)$

2.
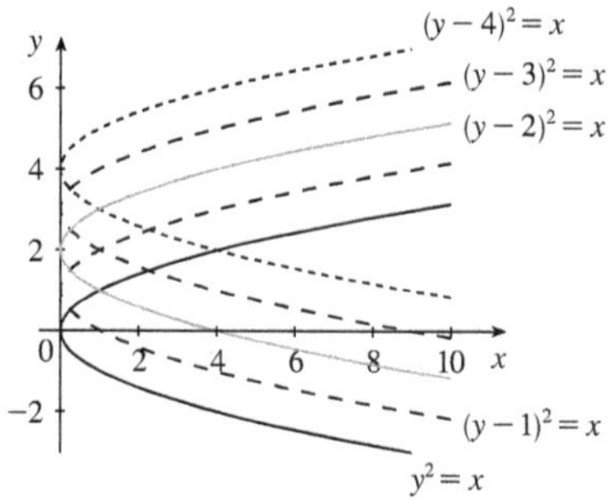

3. $y = \frac{1}{2}x$: x- and y-intercepts $(0, 0)$

$y - 1 = \frac{1}{2}x$: x-intercept $(0, 1)$, y-intercept $(-2, 0)$

$y - 2 = \frac{1}{2}x$: x-intercept $(0, 2)$, y-intercept $(-4, 0)$

$y - 3 = \frac{1}{2}x$: x-intercept $(0, 3)$, y-intercept $(-6, 0)$

4.

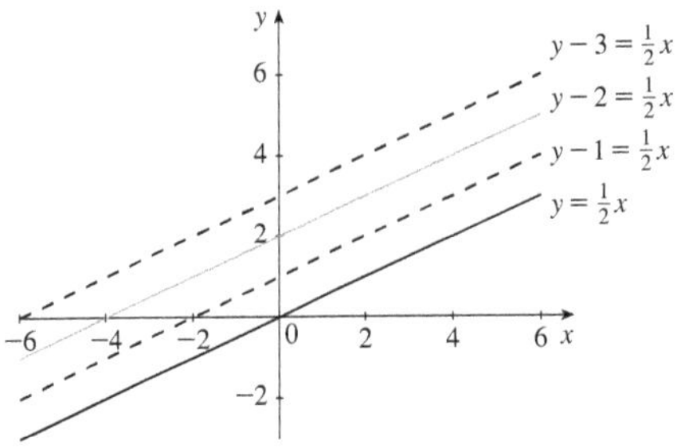

5.

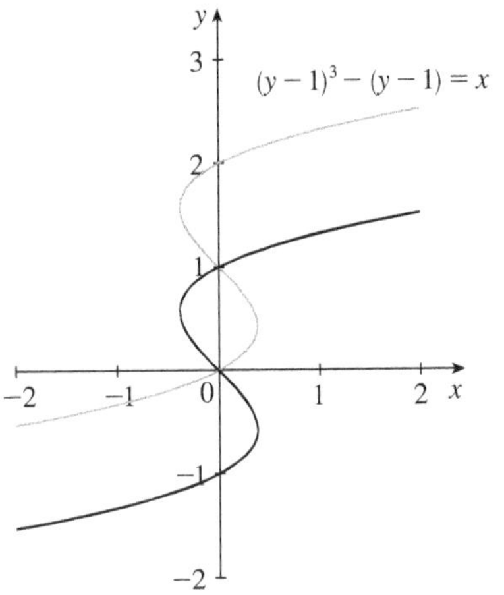

6. Replacing y with $y - k$ shifts a graph k units up.

7. The graph would be shifted 4 units down.

▼ Group Work 3: Symmetry

Don't succumb to the temptation to give hints too early here; this activity should be accessible to any student who has read the material and does not want to give up. If a group finishes early, ask them to write out reasons that certain graphs do not exist.

Answers:

Answers will vary. Problems 5, 6 and 7 are impossible. A straightforward way of showing this is the following pointwise argument: If (x, y) is on the curve, then $(-x, y)$ must also be on it (by symmetry about the x-axis), and then $(x, -y)$ must be on it (by symmetry about the origin), forcing symmetry about the y-axis.

▼ Homework Problems

Core Exercises: 5, 6, 8, 11, 28, 53, 54, 62, 64, 81, 92

Sample Assignment: 5, 6, 8, 11, 14, 27, 28, 53, 54, 62, 64, 75, 81, 92, 94, 105, 115

GROUP WORK 1, SECTION 1.1

Taxicab Geometry

Assume you live in a town where the streets form a grid.

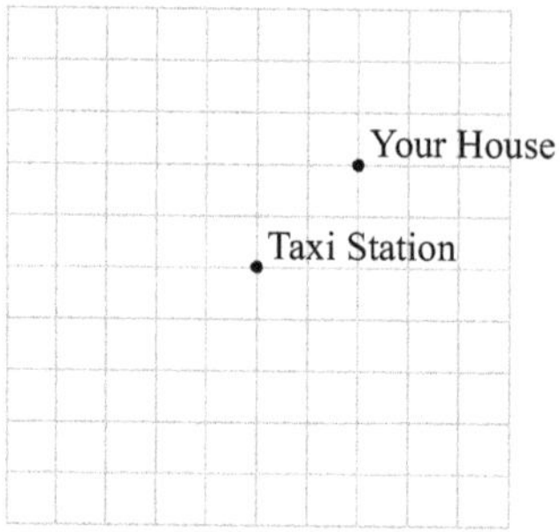

1 unit = 1 mile

1. The taxi dispatcher is willing to let the taxis travel only 3 miles to pick up a customer. Will she allow a taxi to leave her station to pick you up at your house?

2. Shade the region consisting of all points where the cab is allowed to make a pickup.

3. Now assume that, instead of a city, the taxi station and your house are located on a huge expanse of pavement, extending for miles in every direction, with no obstacles in sight. Shade the region where a cab would be allowed to make a pickup.

4. Create a "distance formula" the dispatcher could use to find how far the cab would travel.

GROUP WORK 2, SECTION 1.1

Discovering the Shift

Part One

1. Find the x- and y-intercepts of the curves $y = x^2$, $y = (x-1)^2$, $y = (x-2)^2$, $y = (x-3)^2$, and $y = (x-4)^2$.

2. Sketch the graph of each equation in Problem 1.

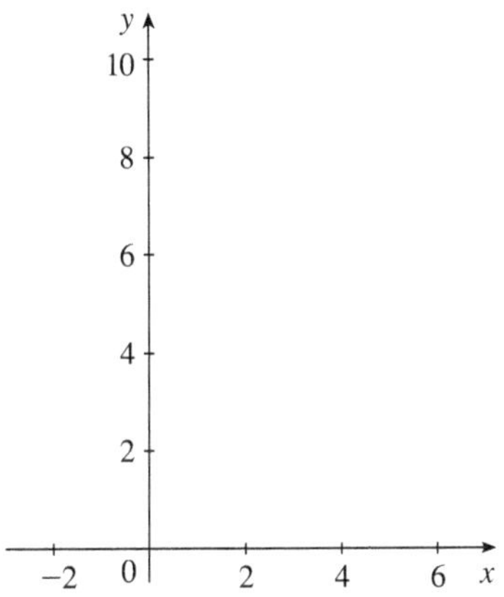

$y = x^2$

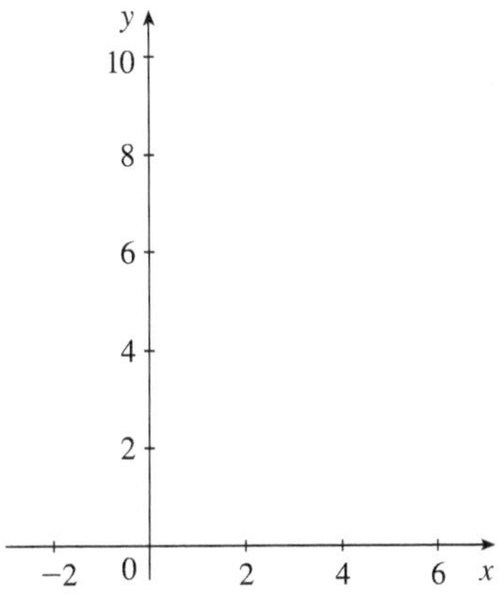

$y = (x-1)^2$

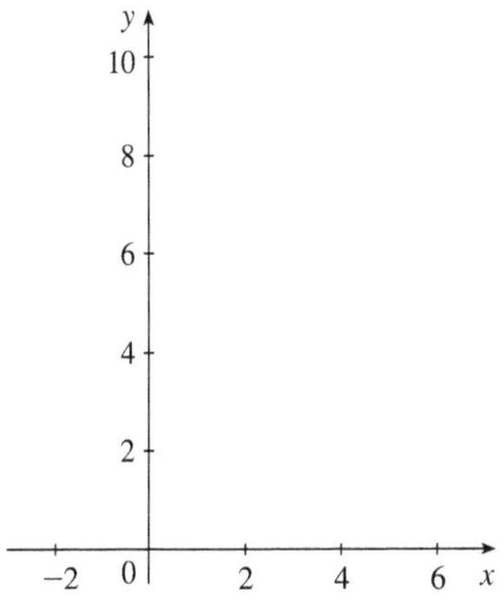

$y = (x-2)^2$

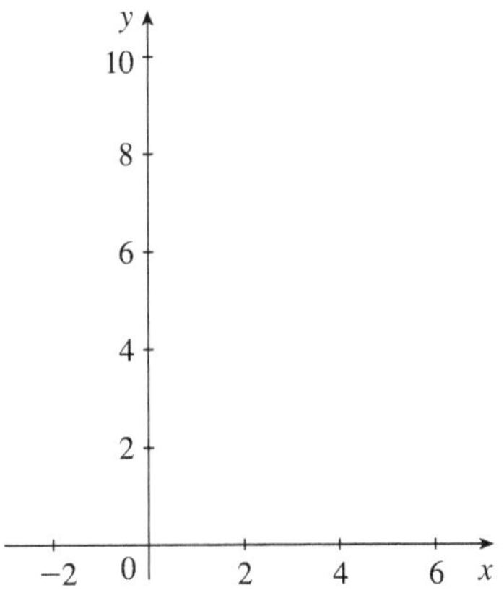

$y = (x-3)^2$

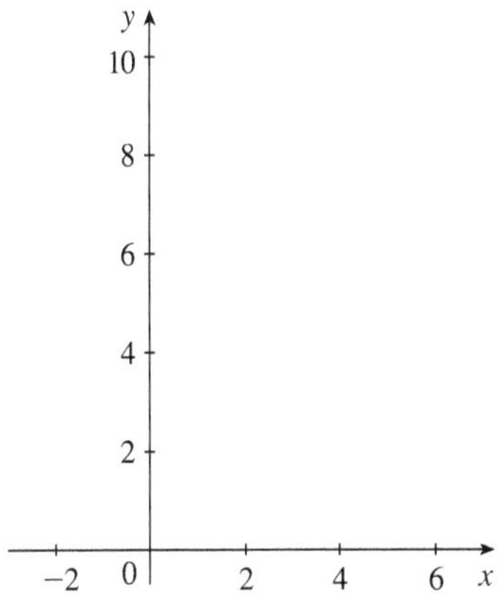

$y = (x-4)^2$

3. Find the x- and y-intercepts of the curves $y = \frac{1}{2}x$, $y = \frac{1}{2}(x - 1)$, $y = \frac{1}{2}(x - 2)$, and $y = \frac{1}{2}(x - 3)$.

4. Sketch the graph of each equation in Problem 3.

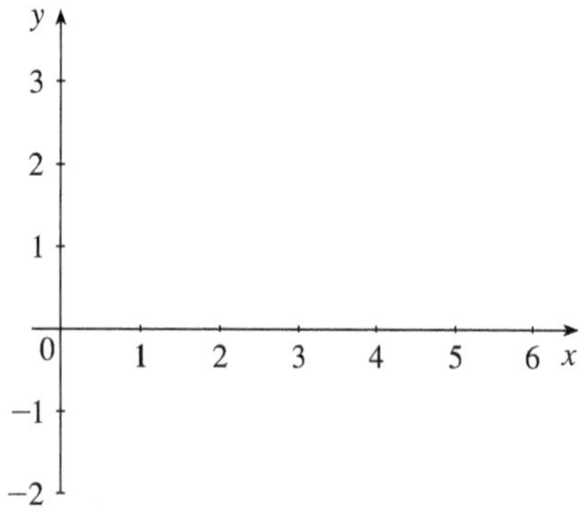

$y = \frac{1}{2}x$

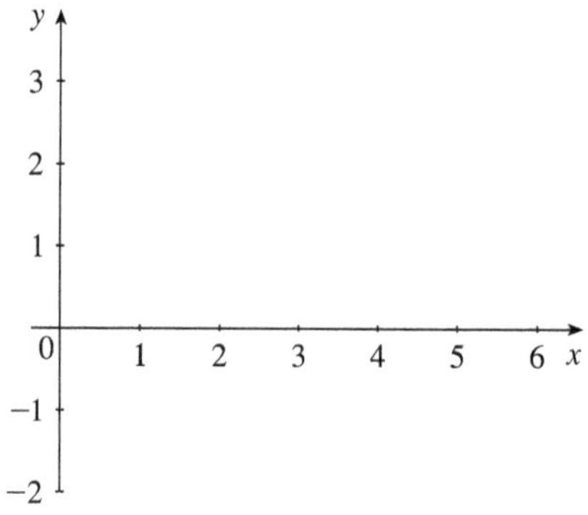

$y = \frac{1}{2}(x - 1)$

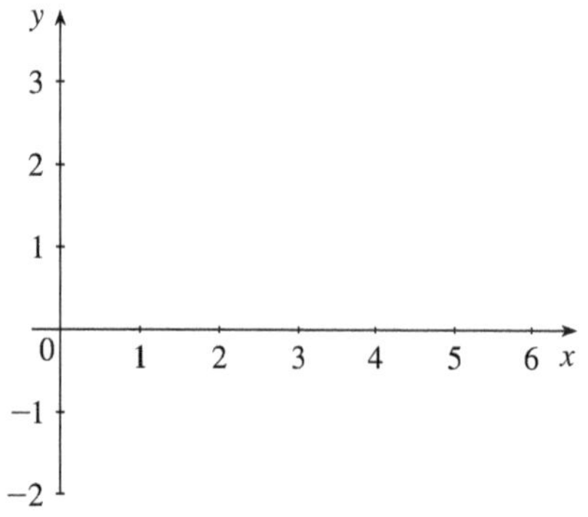

$y = \frac{1}{2}(x - 2)$

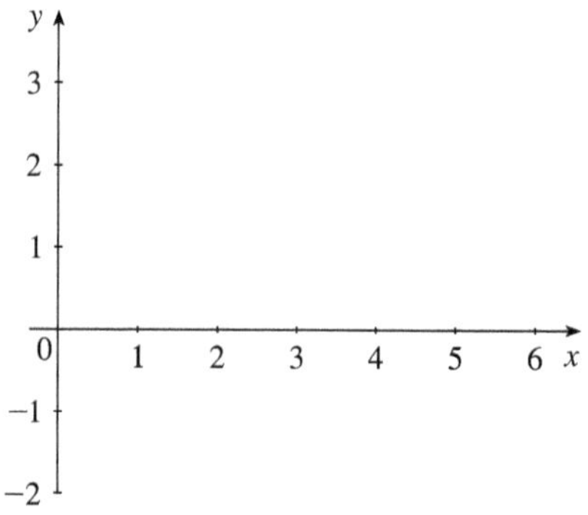

$y = \frac{1}{2}(x - 3)$

5. This is a graph of the complicated equation $y = 2^{\sin x}$.

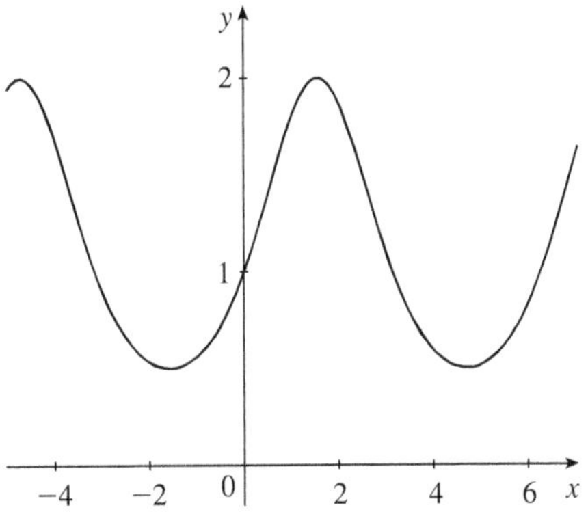

On the same axes, sketch a graph of $y = 2^{\sin(x-2)}$. You don't need to know what "sin" means to do this problem.

6. This activity demonstrates an important principle about replacing x by $x - k$. What is that principle?

Part Two

1. Find the x- and y-intercepts of the curves $y^2 = x$, $(y-1)^2 = x$, $(y-2)^2 = x$, $(y-3)^2 = x$, and $(y-4)^2 = x$.

2. Sketch the graph of each equation in Problem 1.

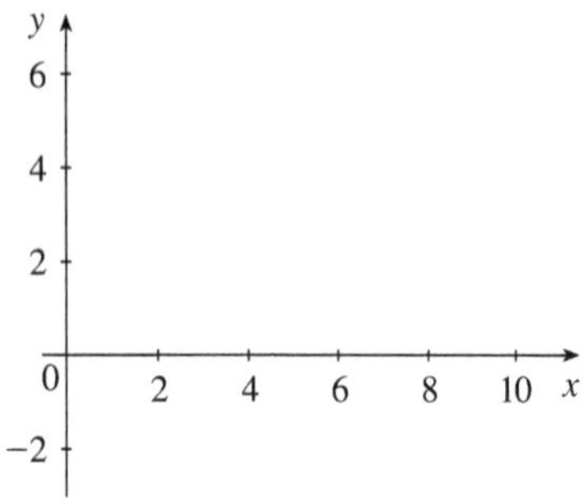

$y^2 = x$

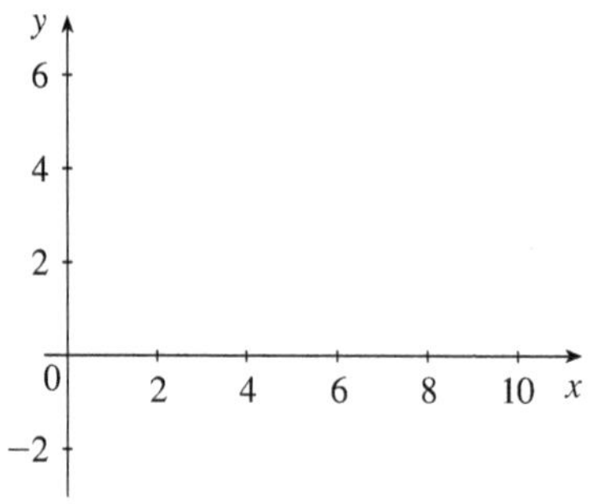

$(y-1)^2 = x$

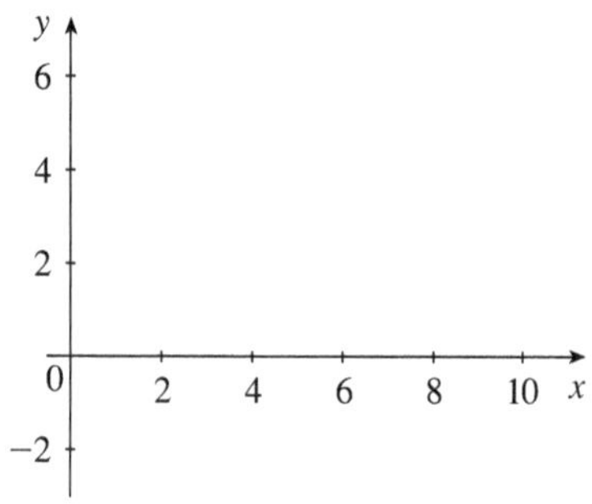

$(y-2)^2 = x$

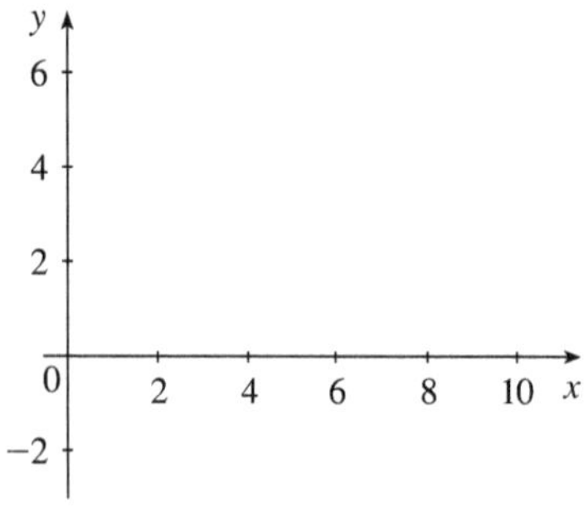

$(y-3)^2 = x$

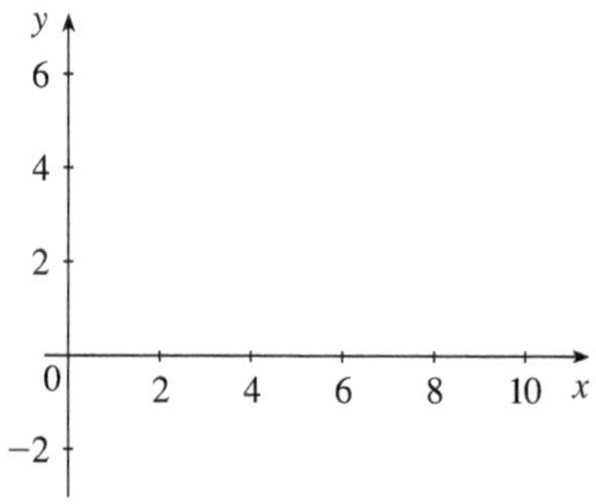

$(y-4)^2 = x$

3. Find the x- and y-intercepts of the curves $y = \frac{1}{2}x$, $y - 1 = \frac{1}{2}x$, $y - 2 = \frac{1}{2}x$, and $y - 3 = \frac{1}{2}x$.

4. Sketch the graph of each equation in Problem 3.

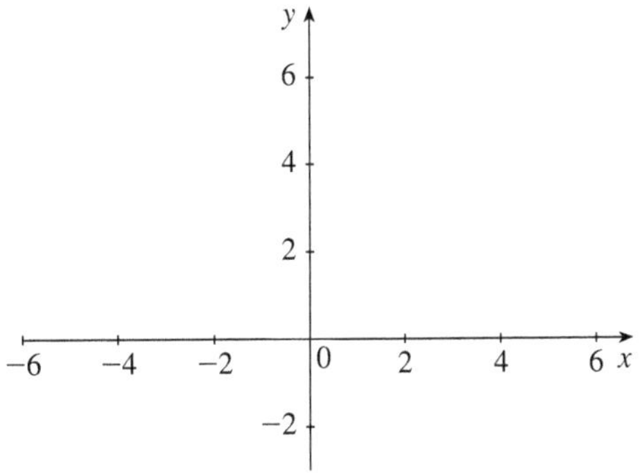

$y = \frac{1}{2}x$

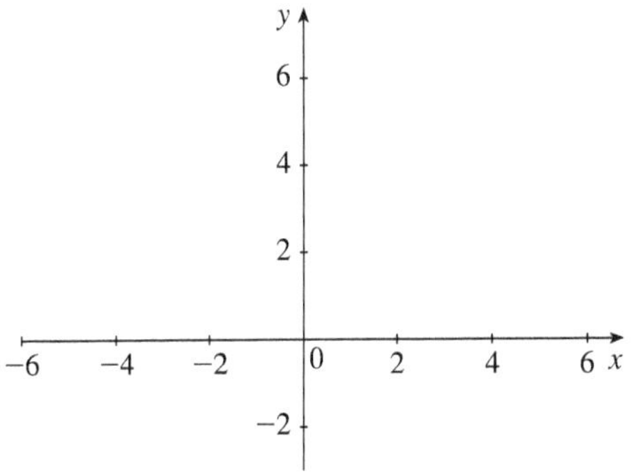

$y - 1 = \frac{1}{2}x$

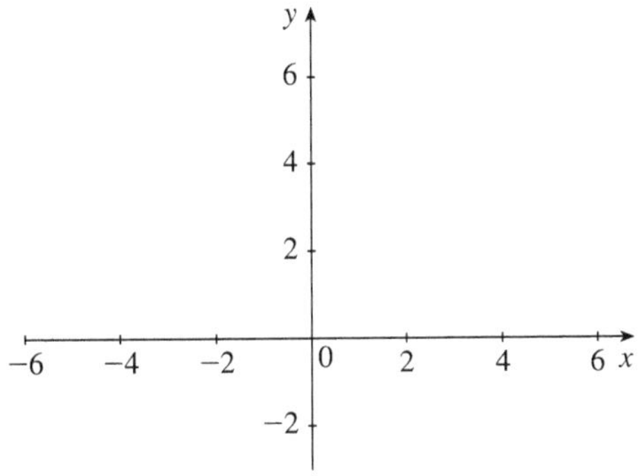

$y - 2 = \frac{1}{2}x$

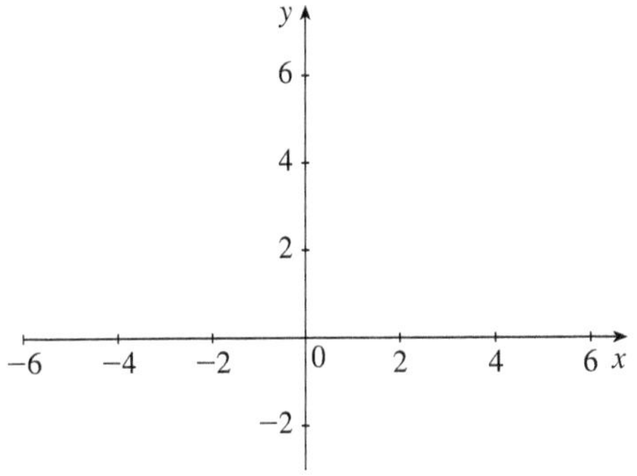

$y - 3 = \frac{1}{2}x$

5. This is a graph of the complicated equation $y^3 - y = x$.

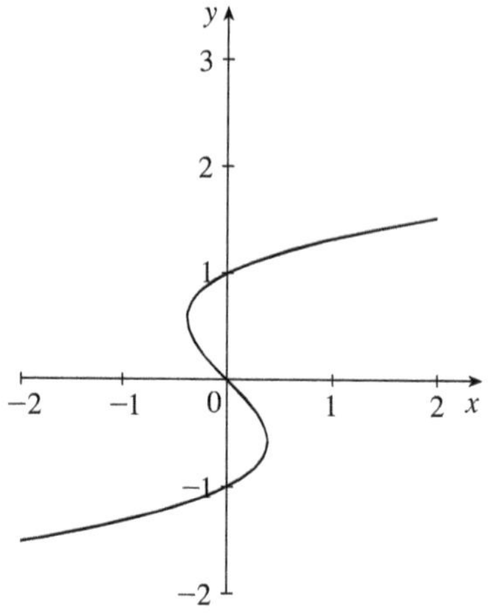

On the same axes, sketch a graph of $(y - 1)^3 - (y - 1) = x$.

6. This activity was designed to demonstrate an important principle about replacing y by $y - k$. What is that principle?

7. What do you think would happen if we replaced y by $y + 4$?

GROUP WORK 3, SECTION 1.1

Symmetry

For this activity you are asked to draw several graphs. If a question asks you to draw something that does not exist, write "Does not exist".

1. Draw a graph that is symmetrical with respect to the x-axis, the y-axis, and the origin.

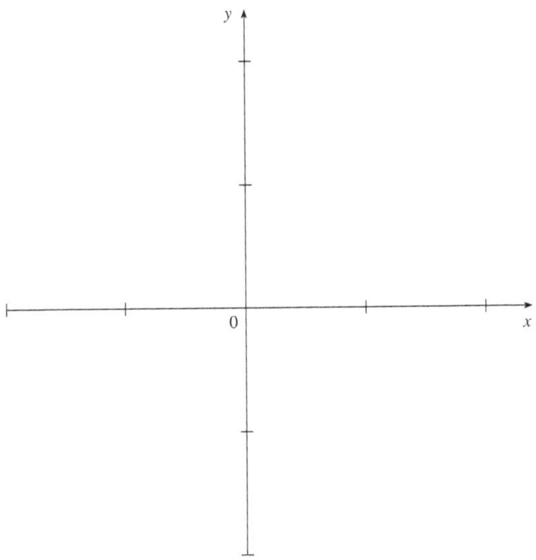

2. Draw a graph that is symmetrical with respect to the x-axis but not the y-axis.

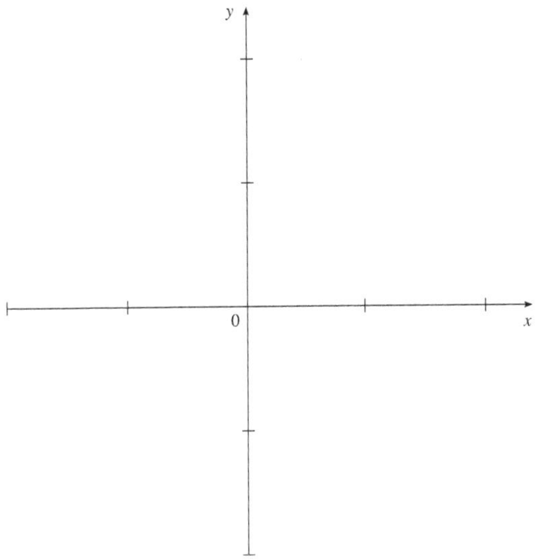

3. Draw a graph that is symmetrical with respect to the y-axis only.

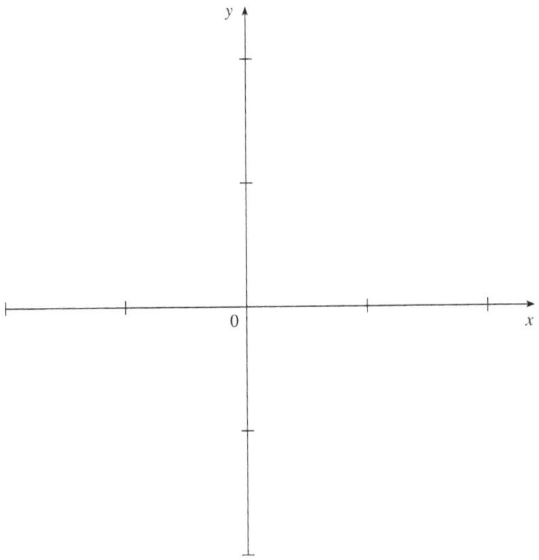

4. Draw a graph that is symmetrical with respect to the origin.

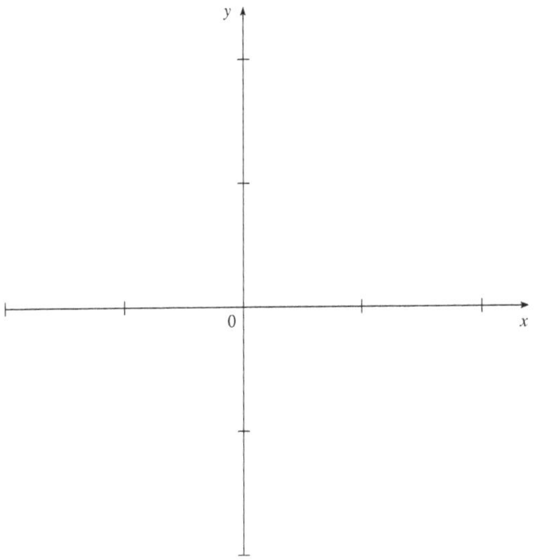

5. Draw a graph that is symmetrical with respect to the x-axis and the y-axis, but not the origin.

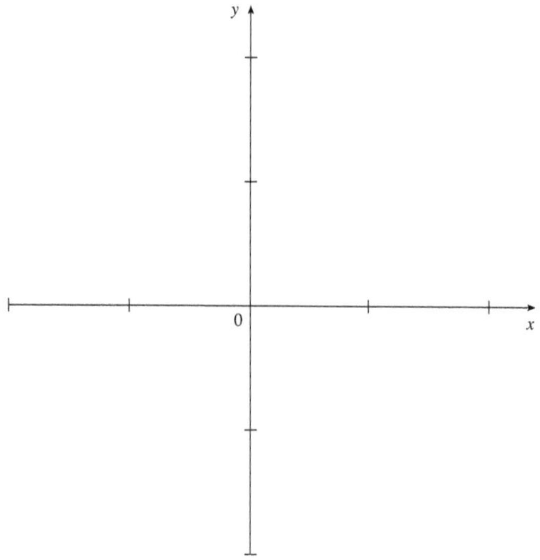

6. Draw a graph that is symmetrical with respect to the x-axis and the origin, but not the y-axis.

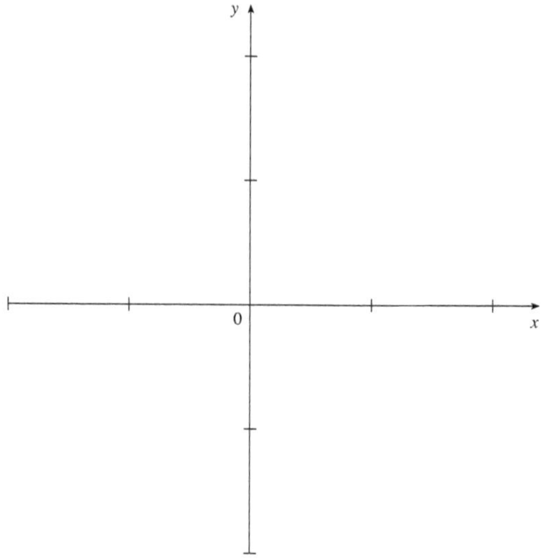

7. Draw a graph that is symmetrical with respect to the y-axis and the origin, but not the x-axis.

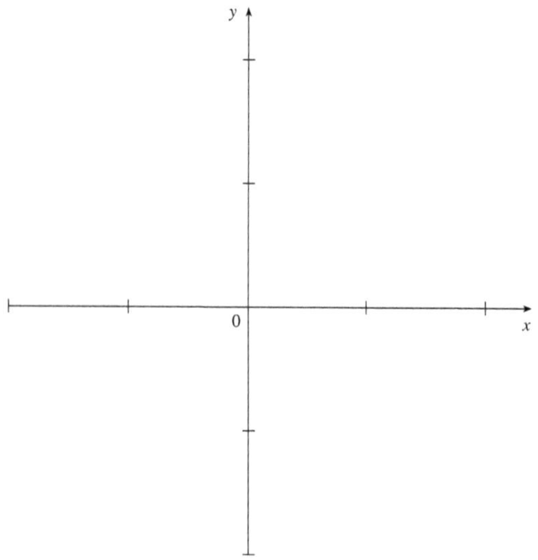

8. Draw a graph that is *not* symmetrical with respect to the x-axis, nor the y-axis, nor the origin.

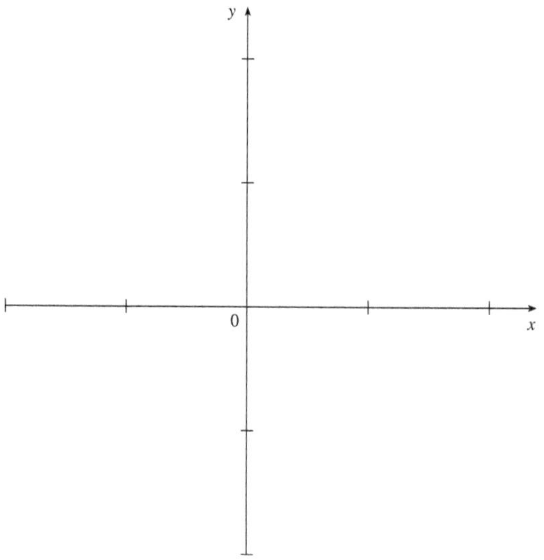

1.2 LINES

▼ Suggested Time and Emphasis

1 class. Essential material.

▼ Points to Stress

1. Computing equations of lines in their various forms, given information such as a point and a slope, or two points.
2. The concepts of parallel, perpendicular, slope, vertical, horizontal, and rate of change as reflected in equations of lines.
3. Graphing lines from their equations.

▼ Sample Questions

- **Text Question:** Which of the three lines graphed below has the largest slope?

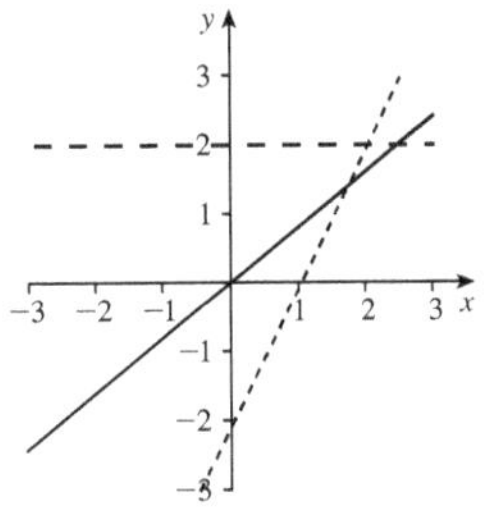

Answer: The dotted line

- **Drill Question:** Find an equation of the line through the points $(2, 4)$ and $(3, -1)$.

Answer: $y = -5x + 14$

▼ In-Class Materials

- One good way to get students used to slope is to have them do quick estimates. Stand in front of the class, raise your arm at various angles, and have them write down (or call out) estimates. Students should be able to tell a positive from a negative slope, and whether a slope is closer to $\frac{1}{3}$ than it is to 3. Also note that if the x- and y-scales on a graph are different, the appearance of the slope can be misleading.
- A classic example is to have the students come up with conversion formulas from Fahrenheit to centigrade and vice-versa, using the fact that in centigrade measurement, water freezes at $0°$ and boils at $100°$ ($-32°$ F and $212°$ F respectively.) Other conversion formulas can be found, using facts such as the following: 3 ft ≈ 0.914 m, 1 L ≈ 0.264 gal (US), a 250-calorie snack is equivalent to 0.991 BTU, etc.
- Go over an interpolation with the students. For example, compare your school's budget this year with that of ten years ago. Find the equation of the appropriate line. Then try to predict the budget five years ago, and see if a linear model was appropriate.
- Ask the students to figure out if three given points are vertices of a right triangle. They can go ahead and plot them, to get a guess going. Ask them to come up with a strategy, and reveal (or ideally elicit) the idea of writing equations of lines between each pair of points, and checking for perpendicular lines. $(3, 4)$, $(3, 12)$, and $(6, 5)$ do not form a right triangle; $(-2, -1)$, $(-2, 8)$, and $(8, -1)$ do.

- Don't neglect to give the students horizontal and vertical lines to explore—they often find equations of vertical lines challenging.

▼ Examples

Equations of some standard lines:

- Through $(3, 5)$ and $(-1, 13)$: $y = -2x + 11$
- Through $(6, -5)$ with a slope of $\frac{1}{3}$: $y = \frac{1}{3}x - 7$

▼ Group Work 1: I've Grown Accustomed to Your Growth

This activity uses the students' skills in finding equations of lines to foreshadow the idea that some growth is linear, and some is not linear.

Answers:

1. Yes ($m = 1$), no, yes ($m \approx 2.08$), yes ($m \approx 2.01$)
2. Equally spaced changes in x-values result in equally spaced changes in y-values.

▼ Group Work 2: Read Between the Lines

This is a drill-oriented activity, probably best worked on by the students individually, using their groups to check their work. It just goes over various bits of information that suffice to give the equation of a line.

Answers:

1. $y = -\frac{15}{2}x + 35$ **2.** $x = 4$ **3.** $y = -2x + 7$ **4.** $y = 3$
5. $x = 2$ **6.** $y = -2x + 7$ **7.** $y = -\frac{1}{3}x + \frac{11}{3}$ **8.** $x = 2$
9. Not enough information given **10.** $y = \frac{1}{2}x$ **11.** $y = 4x - 11$

▼ Group Work 3: Lines and Slopes

In this age of graphing calculators, students usually see graphs where the x- and y-scales are the same. This activity allows them to explore slopes when the scales are different, and allows them to connect linear relationships to the real world.

Answers:

1. (b), (a), (c)
2. (a), (c), (b)
3. Answers will vary. Make sure your students' answers are consistent with the slopes of the lines. Possible answers:
   (a) The accumulation of rain or snow during a storm.
   (b) The distance a jet plane has traveled from a given point in time.
   (c) The force exerted by a spring as it stretches.

▼ Homework Problems

Core Exercises: 1, 2, 7, 16, 21, 24, 26, 37, 46, 52

Sample Assignment: 1, 2, 4, 7, 13, 16, 18, 21, 24, 26, 30, 37, 38, 46, 52, 65, 75

GROUP WORK 1, SECTION 1.2

I've Grown Accustomed to Your Growth

1. Some of the following four tables of data have something in common: linear growth. By trying to find equations of lines, determine which of them represent linear growth. Which table has the slowest growth?

x	y
1	2
2	3
3	4
4	5

x	y
21.5	4.32
32.6	4.203
43.7	4.090
54.8	3.980

x	y
-3	1.1
-2.5	2.14
-2	3.18
-1.5	4.22

x	y
1	-5.00
3	-0.98
6	5.05
8	9.07

2. In a sentence, describe a property of linear growth that can be determined from a table of values.

GROUP WORK 2, SECTION 1.2

Read Between the Lines

Find the equations of the following lines. If there is not enough information given, write "Not enough information given".

1. The line through the points $(4, 5)$ and $(6, -10)$.

2. The line through the points $(4, 5)$ and $(4, -10)$.

3. The line through the point $(2, 3)$ with slope -2.

4. The horizontal line through the point $(2, 3)$.

5. The vertical line through the point $(2, 3)$.

6. The line through the point $(2, 3)$ that is parallel to the line $y = -2x + 11$.

7. The line through the point $(2, 3)$ that is perpendicular to the line $3x - y + 12 = 0$.

8. The line through the point $(2, 3)$ that is perpendicular to the line $y = 5$.

9. The line parallel to the line $y = -2x + 11$ and perpendicular to the line $y = \frac{1}{2}x + 12$.

10. The line with slope $\frac{1}{2}$ that passes through the origin.

11. The line parallel to the line $y = 4x - 2$ that passes through the point of intersection of the lines $y = x - 2$ and $y = 2x - 5$.

GROUP WORK 3, SECTION 1.2

Lines and Slopes

1. Sort the following three lines from smallest slope to largest slope.

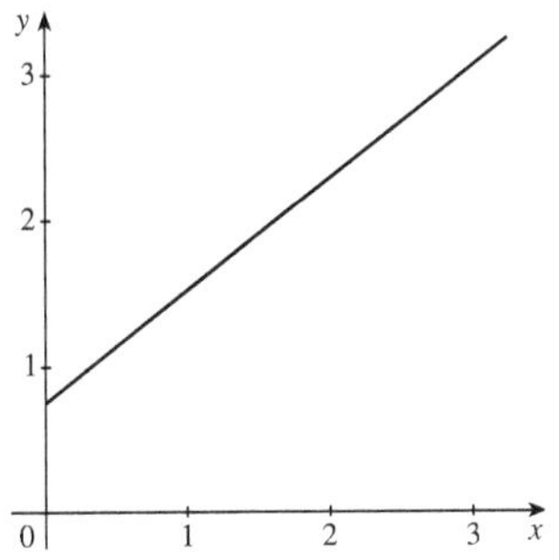

(a)

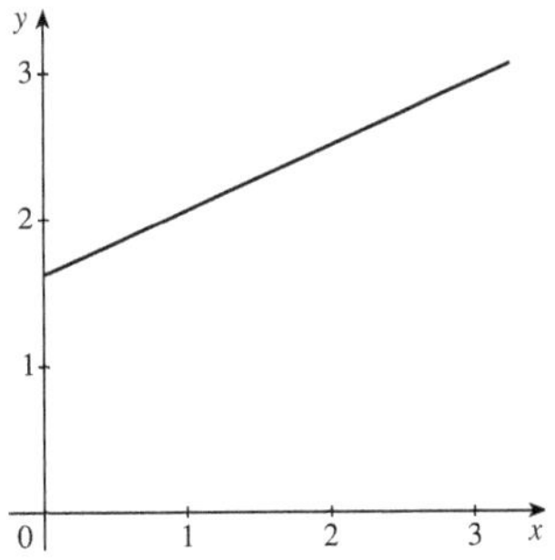

(b)

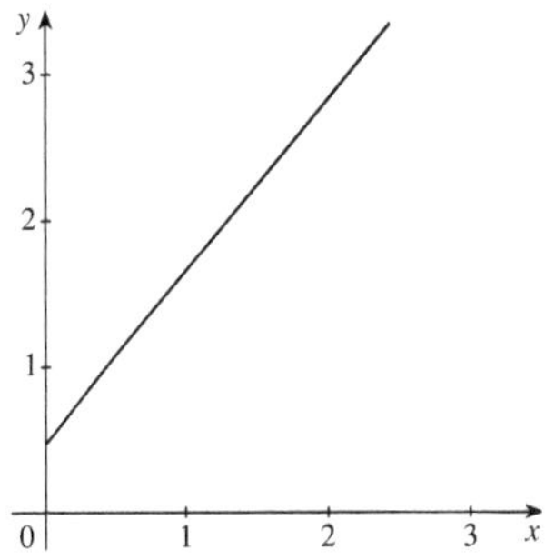

(c)

2. Sort the following three lines from smallest slope to largest slope.

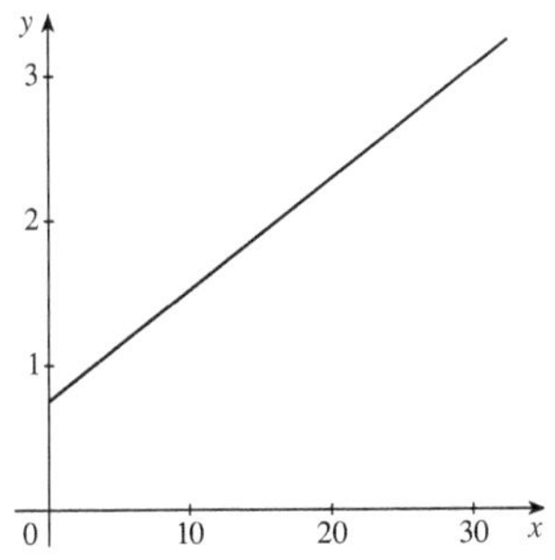

(a)

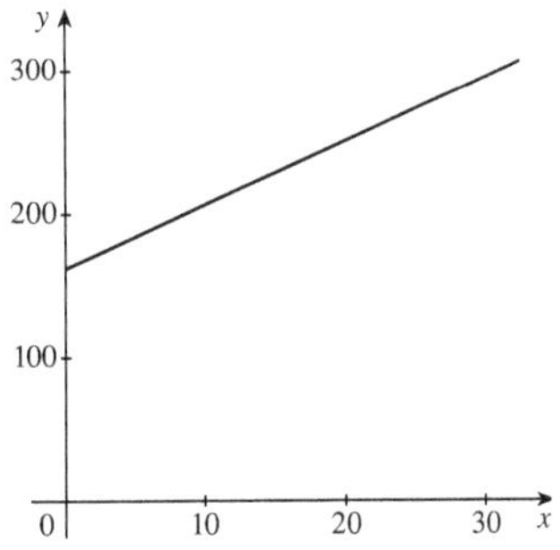

(b)

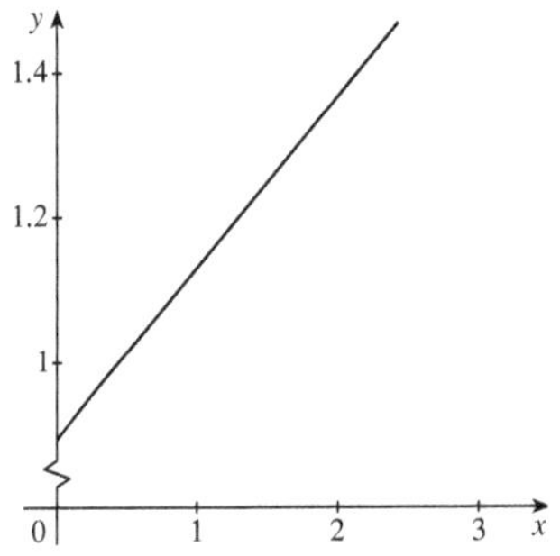

(c)

3. For each of the following graphs, describe a possible real-world situation that it might model. Make sure your answer is realistic.

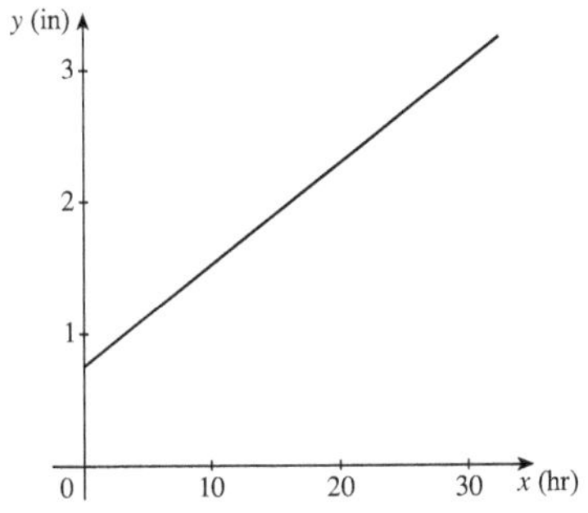

(a)

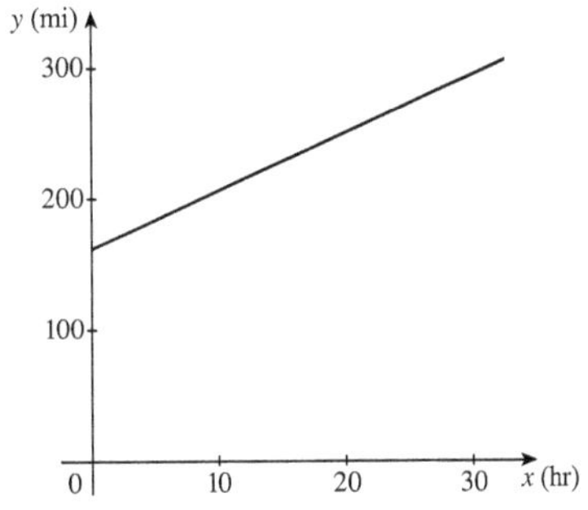

(b)

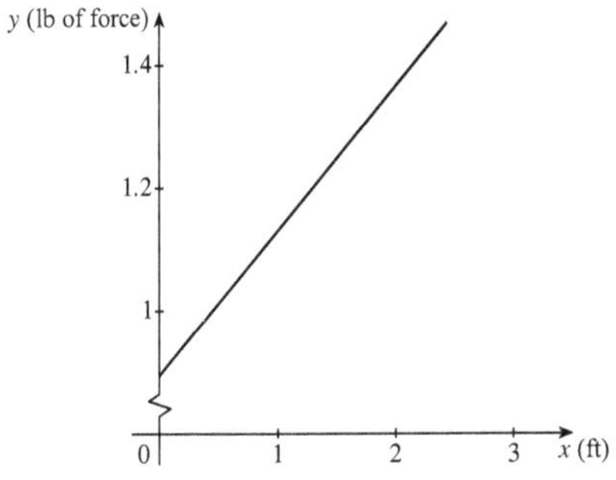

(c)

1.3 WHAT IS A FUNCTION?

▼ Suggested Time and Emphasis

$\frac{1}{2}$–1 class. Essential material.

▼ Points to Stress

1. The idea of function, viewed as the dependence of one quantity on a different quantity.
2. The notation associated with numeric functions, including piecewise-defined functions.
3. Domains and ranges from an algebraic perspective.
4. Four different representations of functions (verbally, algebraically, visually, and numerically).

▼ Sample Questions

- **Text Question:** What is a function?
  Answer: Answers will vary. Anything that gets at the idea of assigning an element in one set to an element in another set should be given full credit.
- **Drill Question:** Let $f(x) = x + \sqrt{x}$. Find $f(0)$ and $f(4)$.
  Answer: $f(0) = 0$, $f(4) = 6$

▼ In-Class Materials

- If students are using calculators, discuss the ties between the idea of a function and a calculator key. Keys such as sin, cos, tan, and $\sqrt{}$ represent functions. It is easy to compute and graph functions on a calculator. Contrast this with equations such as $y^3 - x^3 = 2xy$, which have graphs but are not easy to work with, even with a calculator. (Even symbolic algebra calculators such as the TI-89 do not do well with all general relations.) Point out that the calculator often gives approximations to function values—applying the square root function key to the number 2 gives 1.4142136 which is close, but not equal, to $\sqrt{2}$.
- This course emphasizes functions where both the domain and range sets are numerical. One could give a more abstract definition of function, where D and R can be any set. For example, there is a function mapping each student in the class to his or her birthplace. A nice thing about this point of view is that it can be pointed out that the map from each student to his or her telephone number may *not* be a function, because a student may have more than one telephone number, or none at all.
- Function notation can trip students up. Start with a function such as $f(x) = x^2 - x$ and have your students find $f(0)$, $f(1)$, $f(\sqrt{3})$, and $f(-1)$. Then have them find $f(\pi)$, $f(y)$, and (of course) $f(x+h)$. Some students will invariably, some day, assume that $f(a+b) = f(a) + f(b)$ for all functions, but this can be minimized if plenty of examples such as $f(2+3)$ are done at the outset.

- Discuss the straightforward things to look for when trying to find the domain of a function: zero denominators and negative even roots.

  Discuss the domain and range of a function such as $f(x) = \begin{cases} x^2 & \text{if } x \text{ is an integer} \\ 0 & \text{if } x \text{ is not an integer} \end{cases}$

  If the class seems interested, perhaps let them think about $f(x) = \begin{cases} x^2 & \text{if } x \text{ is rational} \\ 0 & \text{if } x \text{ is irrational} \end{cases}$

- Let $f(x) = \dfrac{x(x-2)}{x-2}$ and $g(x) = x$. Ask students if the functions are the same function. If they say "yes", ask them to compare the domains, or to compute $g(2)$ and $f(2)$. If they say "no", ask them to find a value such that $f(x) \neq g(x)$. [This activity assumes that students know the equation of a circle with radius r. If they do not, this may be a good opportunity to introduce the concept.]

▼ Examples

- Real-world piecewise functions:
  1. The cost of mailing a parcel that weighs w ounces (see Figure 1 in the text)
  2. The cost of making x photocopies (given that there is usually a bulk discount)
  3. The cost of printing x pages from a computer (at some point the toner cartridge must be replaced)
- A function with a nontrivial domain: $\sqrt{\dfrac{x^2-5x+6}{x^2-2x+1}}$ has domain $(-\infty, 1) \cup (1, 2] \cup [3, \infty)$.

▼ Group Work 1: Choosing a Calling Plan

This activity will require some advance work on the part of your students, but it will be worth it. It complements Example 3 in the text. The world is full of advertisements for long-distance services, phone plans, and the like. These often have figures such as "five cents per minute", but there are a lot of details behind the figures. There may or may not be a monthly fee. There may or may not be a minimum call length. (One commercial offered a 20-minute call for $1, but did not mention that a three-minute call also cost $1.) There may or may not be a fee change after a certain time. (One plan costs five cents per minute for the first twenty minutes and seven cents per minute thereafter.) There may also be a rate increase after a certain base number of minutes are used.

Have each student pick a plan (or you can assign plans to them) and research exactly how the plan works. In addition to the myriad phone and long distance plans, there are collect call plans, plans in which phone service is bundled with other services, etc. Students can also research the cost of an operator-assisted long distance call. After they have done so, this group work can be handed out.

Part 1 is for this section and is algebraic in nature. Part 2 is for the next section (although it can be done earlier) and involves drawing a graph of a piecewise function. Note that you may not want to discuss the greatest integer function at this point. In that case, you can let students assume that the call lengths are all an integral number of minutes.

You may have to "keep students honest" here—a careless student may take the phrase "five cents per minute" at face value and not worry about the fine print. Being able to listen carefully to a commercial or a salesperson and translate the pitch into mathematics is not a trivial skill.

It might be a good idea to collect students' findings and distribute them. Students can then be assigned to decide which plan is best for them, taking into account their average call length, and the amount of calling they do each month. (The monthly fees become important to a person who doesn't use the phone often.) An alternative form of this group work is provided that deals with text messages.

▼ Group Work 2: Finding a Formula

Make sure that students know the equation of a circle with radius r, and that they remember the notation for piecewise-defined functions. Divide the class into groups of four. In each group, have half of them work on each problem first, and then have them check each other's work. If students find these problems difficult, have them work together on each problem.

Answers: **1.** $f(x) = \begin{cases} -x-2 & \text{if } x \le -2 \\ x+2 & \text{if } -2 < x \le 0 \\ 2 & \text{if } x > 0 \end{cases}$ **2.** $g(x) = \begin{cases} x+4 & \text{if } x \le -2 \\ 2 & \text{if } -2 < x \le 0 \\ \sqrt{4-x^2} & \text{if } 0 < x \le 2 \\ x-2 & \text{if } x > 2 \end{cases}$

▼ Group Work 3: Rounding the Bases

On the board, review how to compute the percentage error when estimating π by $\frac{22}{7}$. (Answer: 0.04%) Have them work on the problem in groups. If a group finishes early, have them look at $h(7)$ and $h(10)$ to see how fast the error grows. Students have not seen exponential functions before, but Problem 3 is a good foreshadowing of Section 8.1.

Answers: **1.** 17.811434627, 17, 4.56% **2.** 220.08649875, 201, 8.67% **3.** 45.4314240633, 32, 29.56%

▼ Homework Problems

Core Exercises: 2, 6, 11, 18, 21, 29, 32, 44, 47, 50, 55, 65

Sample Assignment: 2, 6, 7, 11, 18, 21, 22, 25, 29, 32, 36, 44, 47, 50, 55, 56, 63, 65, 72, 83

GROUP WORK 1, SECTION 1.3

Choosing a Calling Plan

1. You have been assigned to research a calling plan. What is the name of the plan you've investigated?

2. Is there a monthly fee?

For Questions 3—7, do not include monthly fees.

3. How much would it cost to make a twenty-minute call using your plan?

4. How much would it cost to make a five-minute call?

5. How much would it cost to make a one-minute call?

6. How much would it cost to make a three-hour call?

7. Write a function $c(t)$, where t is the duration of the call in minutes and $c(t)$ is the cost of the call.

8. How much would it cost to make 20 five-minute calls in a month? (Include the monthly fee.)

9. How much would it cost to make 2 fifty-minute calls in a month? (Include the monthly fee.)

10. Is this plan suitable for you? Why or why not?

GROUP WORK 1, SECTION 1.3

Choosing a Text Message Plan

1. You have been assigned to research a text message plan. What is the name of the plan you've investigated?

2. Is there a monthly fee?

For Questions 3–5, do not include monthly fees.

3. How much would it cost to send twenty messages in one month using your plan?

4. How much would it cost to send three hundred messages in one month?

5. How much would it cost to send just one message in one month?

6. Is there a maximum message length?

7. Now including the monthly fee, how much would it cost to send 20 messages in one month. 300 messages? A single message?

8. What is the average cost per message under this plan if you send 20 messages in one month. 300 message? A single message?

9. Write a function $c(n)$ modeling the total monthly cost c if n messages are sent that month.

10. Is this plan suitable for you? Why or why not?

GROUP WORK 2, SECTION 1.3

Finding a Formula

Find formulas for the following functions:

1.

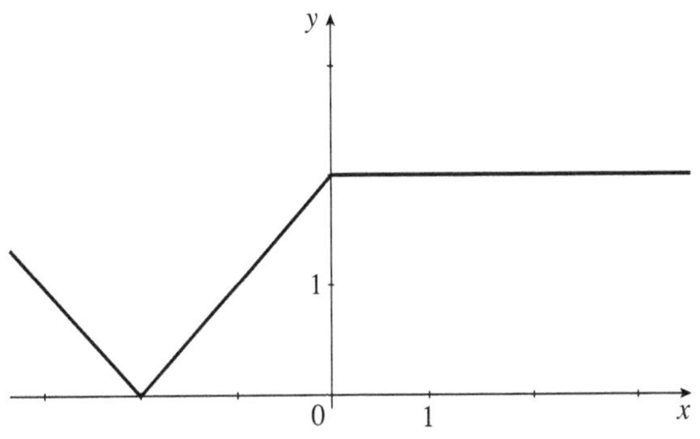

2.

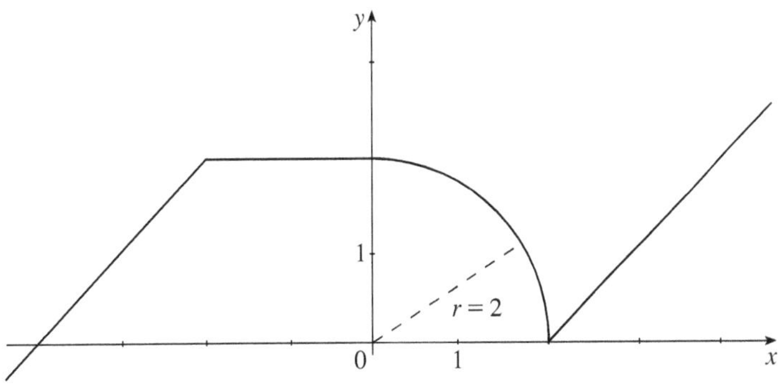

GROUP WORK 3, SECTION 1.3

Rounding the Bases

1. For computational efficiency and speed, we often round off constants in equations. For example, consider the linear function

$$f(x) = 3.137619523x + 2.123337012$$

In theory, it is easy and quick to find $f(1)$, $f(2)$, $f(3)$, $f(4)$, and $f(5)$. In practice, most people doing this computation would probably substitute

$$f(x) = 3x + 2$$

unless a very accurate answer is called for. For example, compute $f(5)$ both ways to see the difference.

The actual value of $f(5)$: __________

The "rounding" estimate: __________

The percentage error: __________

2. Now consider

$$g(x) = 1.12755319x^3 + 3.125694x^2 + 1$$

Again, one is tempted to substitute $g(x) = x^3 + 3x^2 + 1$.

The actual value of $g(5)$: __________

The "rounding" estimate: __________

The percentage error: __________

3. It turns out to be dangerous to similarly round off exponential functions, due to the nature of their growth. For example, let's look at the function

$$h(x) = (2.145217198123)^x$$

One may be tempted to substitute $h(x) = 2^x$ for this one. Once again, look at the difference between these two functions.

The actual value of $h(5)$: __________

The "rounding" estimate: __________

The percentage error: __________

1.4 GRAPHS OF FUNCTIONS

▼ Suggested Time and Emphasis

1 class. Essential material.

▼ Points to Stress

1. The Vertical Line Test.
2. Graphs of piecewise-defined functions.
3. The greatest integer function.

▼ Sample Questions

- **Text Question:** Your text discusses the greatest integer function $[\![x]\!]$. Compute $[\![2.6]\!]$, $[\![2]\!]$, $[\![-2.6]\!]$, and $[\![-2]\!]$.
  Answer: $[\![2.6]\!] = 2$, $[\![2]\!] = 2$, $[\![-2.6]\!] = -3$, $[\![-2]\!] = -2$
- **Drill Question:** Let $f(x) = x^2 + |x|$. Which of the following is the graph of f? How do you know?

  (a)
  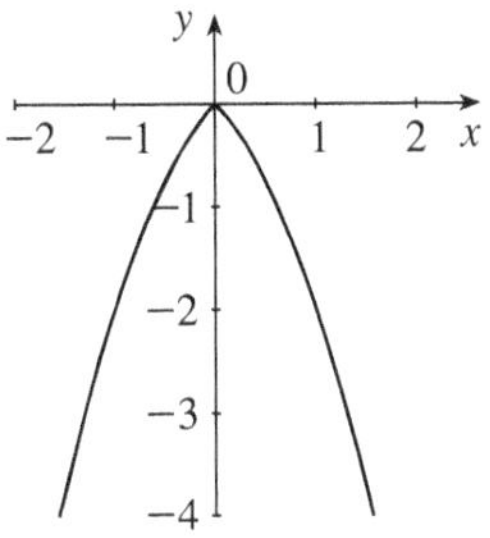
  (b)
  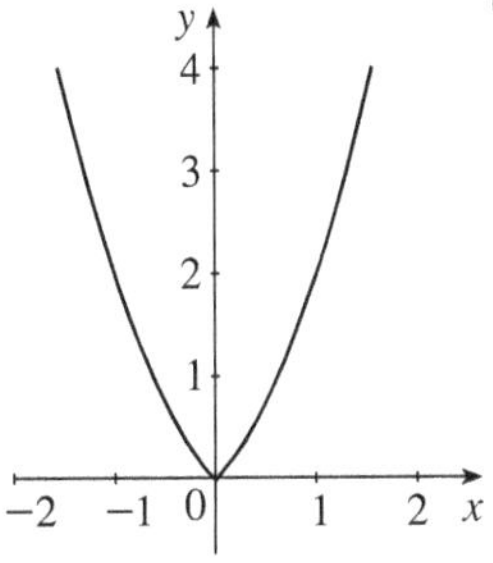
  (c)
  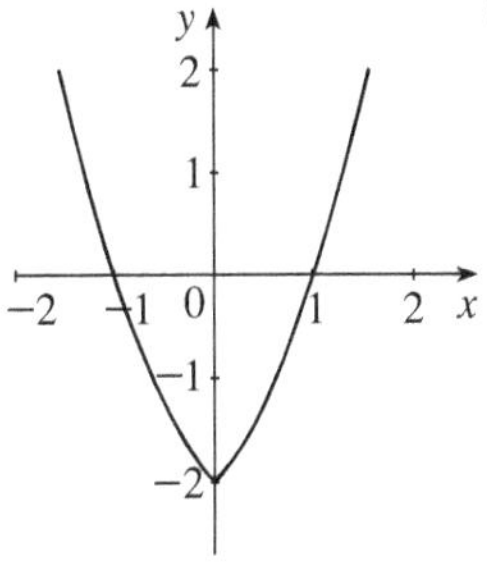
  (d)
  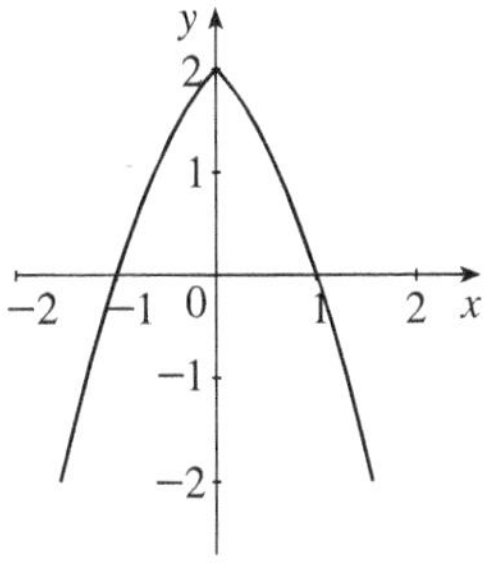

  Answer: (b) is the graph of f, because $f(x) \geq 0$ for all x.

▼ In-Class Materials

- Draw a graph of fuel efficiency versus time on a trip, such as the one below. Lead a discussion of what could have happened on the trip.

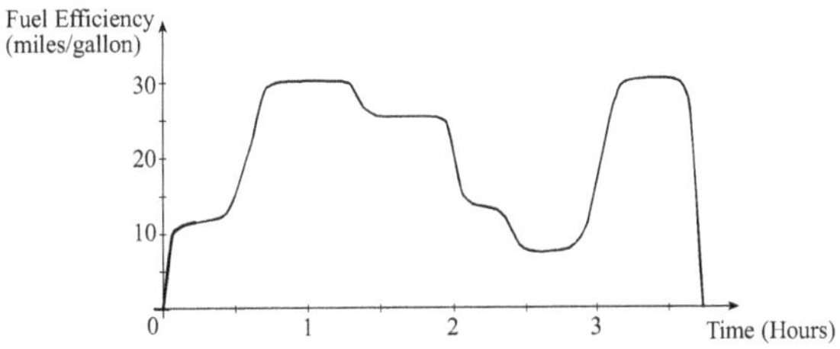

- In 1984, United States President Ronald Reagan proposed a plan to change the personal income tax system. According to his plan, the income tax would be 15% on the first \$19,300 earned, 25% on the next \$18,800, and 35% on all income above and beyond that. Describe this situation to the class, and have them graph (marginal) tax rate and tax owed versus income for incomes ranging from \$0 to \$80,000. Then have them try to come up with equations describing this situation.
- In the year 2000, Presidential candidate Steve Forbes proposed a "flat tax" model: 0% on the first \$36,000 and 17% on the rest. Have your students do the same analysis, and compare the two models. As an extension, consider having them look at a current tax table and draw similar graphs.

- Discuss the shape, symmetries, and general "flatness" near 0 of the power functions x^n for various values of n. Similarly discuss $\sqrt[n]{x}$ for n even and n odd. A blackline master is provided at the end of this section, before the group work handouts.

▼ Examples

- A continuous piecewise-defined function

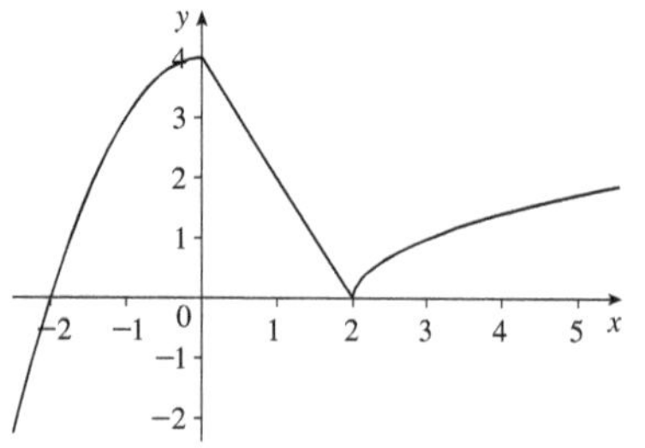

$$f(x) = \begin{cases} 4 - x^2 & \text{if } x < 0 \\ 4 - 2x & \text{if } 0 \le x \le 2 \\ \sqrt{x - 2} & \text{if } x > 2 \end{cases}$$

- A discontinuous piecewise-defined function

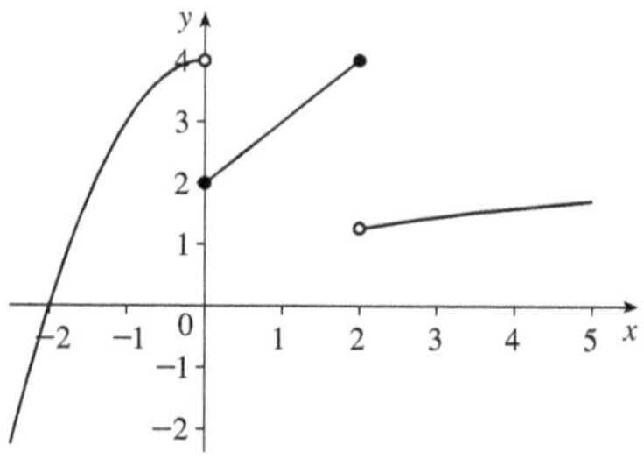

$$f(x) = \begin{cases} 4 - x^2 & \text{if } x < 0 \\ x + 2 & \text{if } 0 \le x \le 2 \\ \sqrt[3]{x} & \text{if } x > 2 \end{cases}$$

- Classic rational functions with interesting graphs

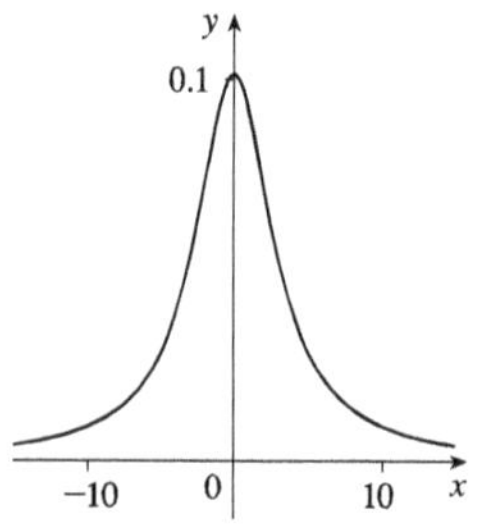

$$h(x) = \frac{1}{x^2 + \pi^2}$$

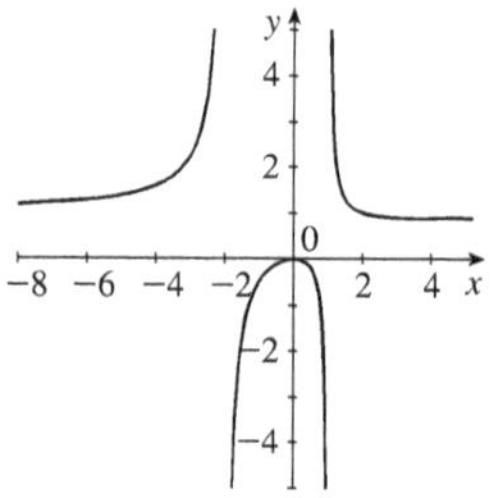

$$i(x) = \frac{x^2}{x^2 + x - 2}$$

▼ Group Work 1: Every Picture Tells a Story

Put students in groups of four, and have them work on the exercise. If there are questions, encourage them to ask each other before asking you. After going through the correct matching with them, have each group tell their story to the class and see if it fits the remaining graph.

Answers:

1. (b) **2.** (a) **3.** (c)

4. The roast was cooked in the morning and put in the refrigerator in the afternoon.

▼ Group Work 2: Functions in the Classroom

Before starting this one, review the definition of "function". Some of the problems can be answered only by polling the class after they are finished working. Don't forget to take leap years into account for the eighth problem. For an advanced class, anticipate Section 2.8 by quickly defining "one-to-one" and "bijection", then determining which of the functions have these properties.

Answers:

Chairs: Function, one-to-one, bijection (if all chairs are occupied)

Eye color: Function, not one-to-one

Mom & Dad's birthplace: Not a function; mom and dad could have been born in different places

Molecules: Function, one-to-one (with nearly 100% probability); inverse assigns a number of molecules to the appropriate student.

Spleens: Function, one-to-one, bijection. Inverse assigns each spleen to its owner.

Pencils: Not a function; some people may have more than one or (horrors!) none.

Social Security Number: Function, one-to-one; inverse assigns each number to its owner.

February birthday: Not a function; not defined for someone born on February 29.

Birthday: Function, perhaps one-to-one.

Cars: Not a function; some have none, some have more than one.

Cash: Function, perhaps one-to-one.

Middle names: Not a function; some have none, some have more than one.

Identity: Function, one-to-one, bijection. Inverse is the same as the function.

Algebra instructor: Function, not one-to-one.

▼ Group Work 3: Rational Functions

Remind students of the definition of a rational function as a quotient of polynomials. Students should be able to do this activity by plotting points and looking at domains and ranges.

Answers:

1.

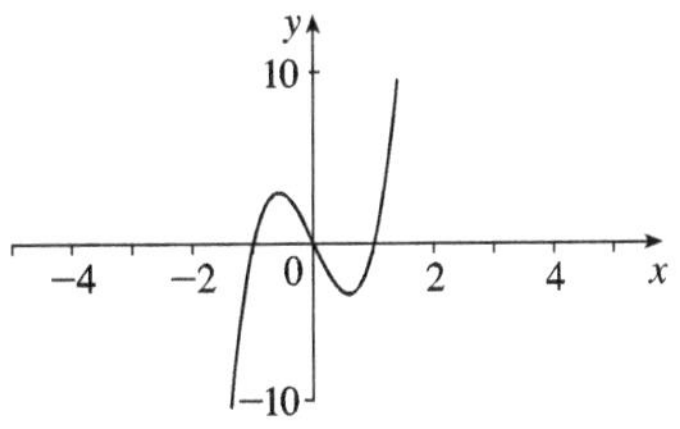

2. (d) **3.** (b) **4.** (e) **5.** (c) **6.** (a)

▼ Homework Problems

Core Exercises: 4, 7, 12, 16, 21, 38, 49, 52, 75

Sample Assignment: 4, 7, 12, 16, 21, 27, 30, 38, 44, 49, 52, 53, 63, 70, 75, 77, 84

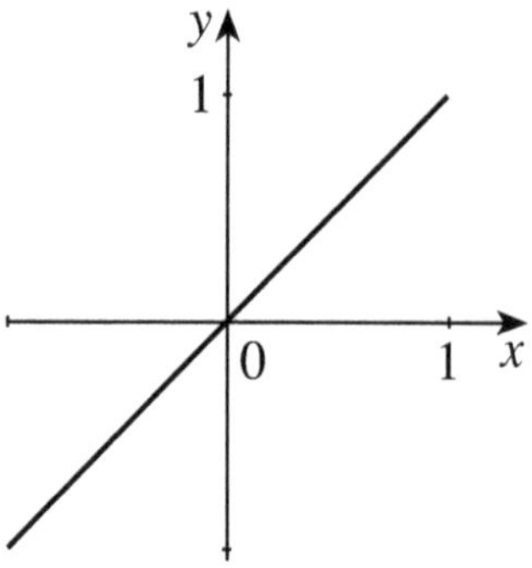

x

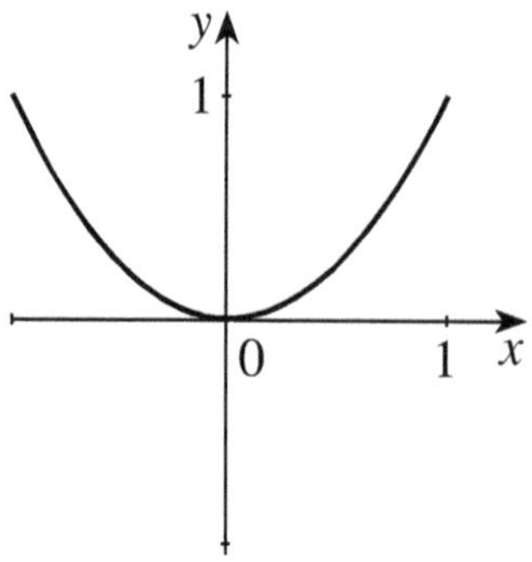

x^2

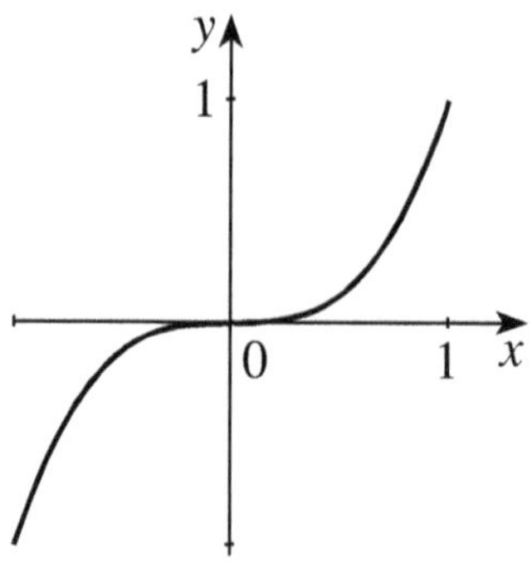

x^3

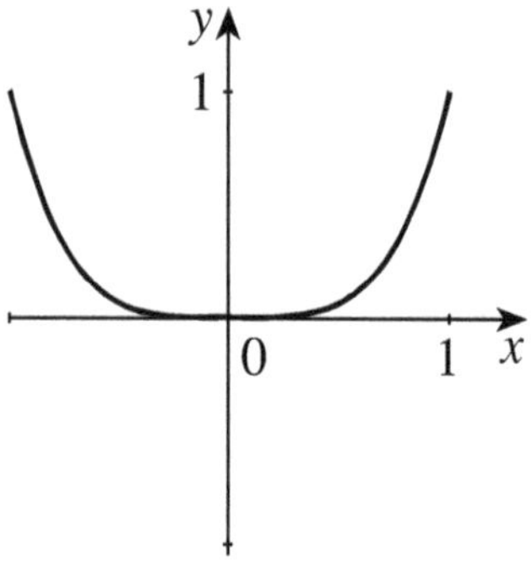

x^4

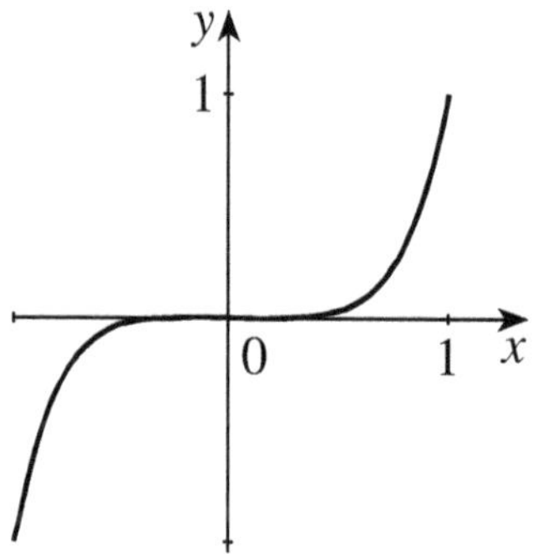

x^5

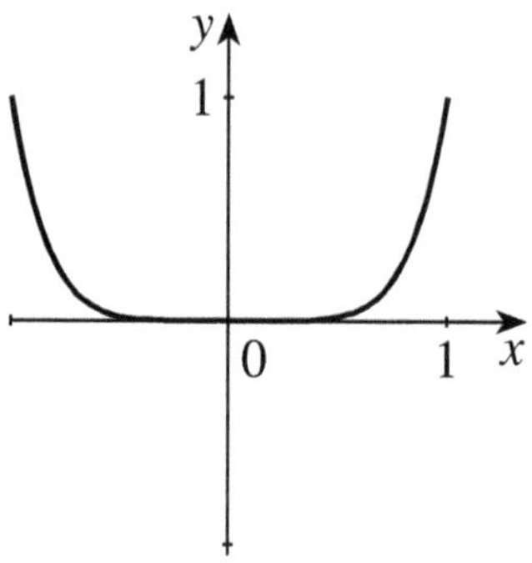

x^6

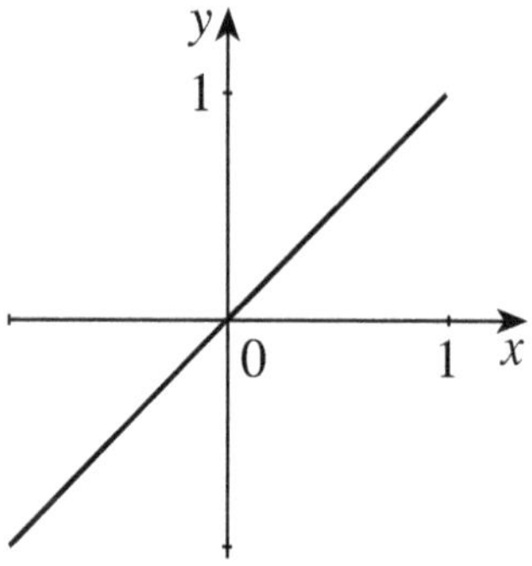

x

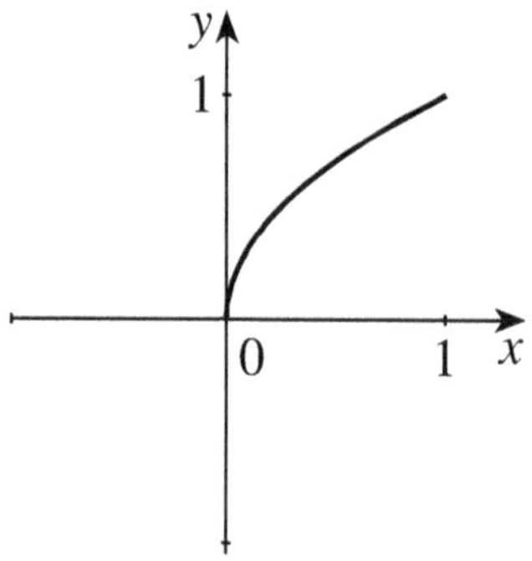

$\sqrt{x}$

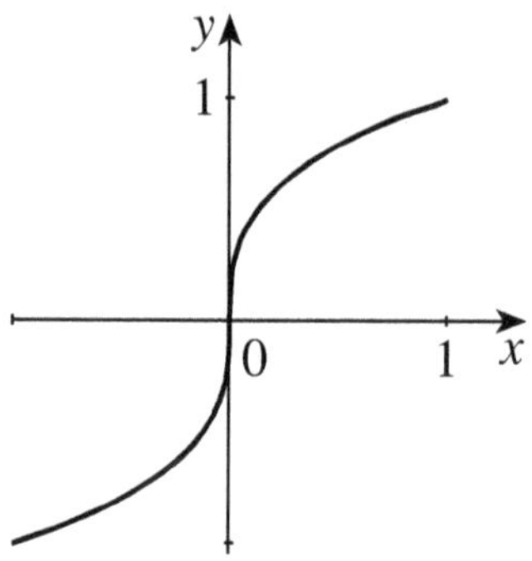

$\sqrt[3]{x}$

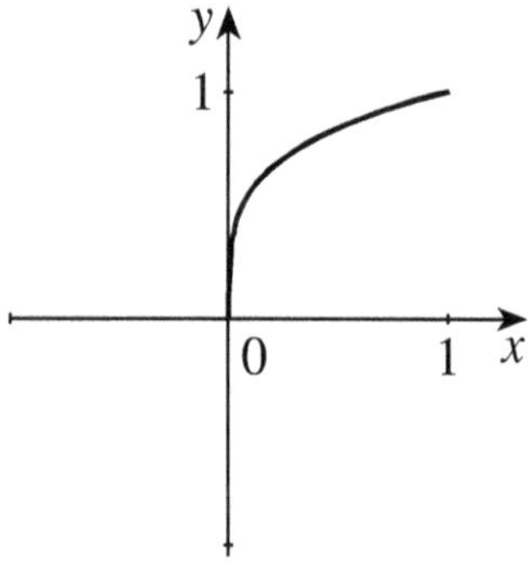

$\sqrt[4]{x}$

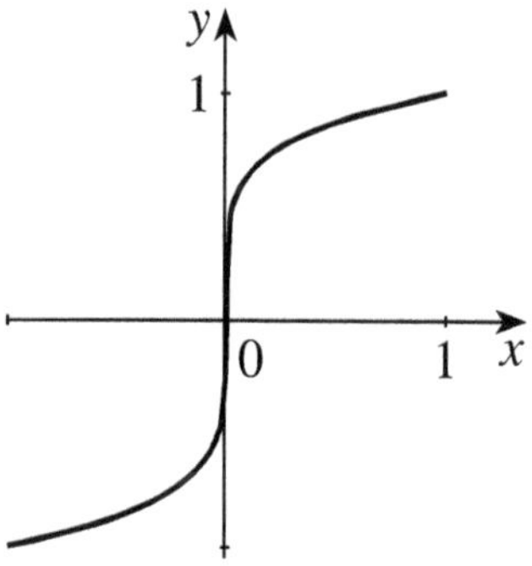

$\sqrt[5]{x}$

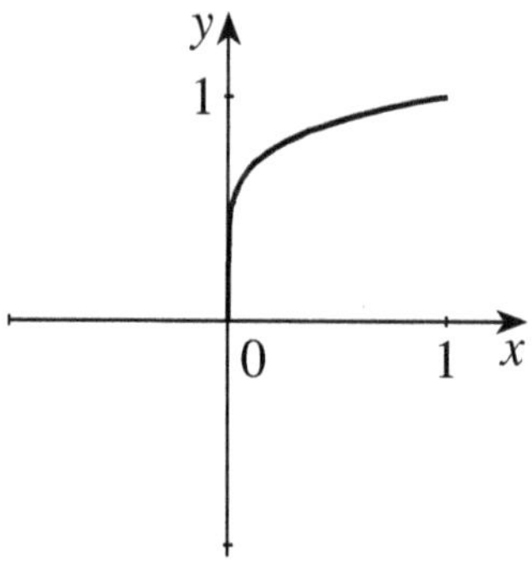

$\sqrt[6]{x}$

GROUP WORK 1, SECTION 1.4

Every Picture Tells a Story

One of the skills you will be learning in this course is the ability to take a description of a real-world occurrence, and translate it into mathematics. Conversely, given a mathematical description of a phenomenon, you will learn how to describe what is happening in plain language. Here follow four graphs of temperature versus time and three stories. Match the stories with the graphs. When finished, write a similar story that would correspond to the final graph.

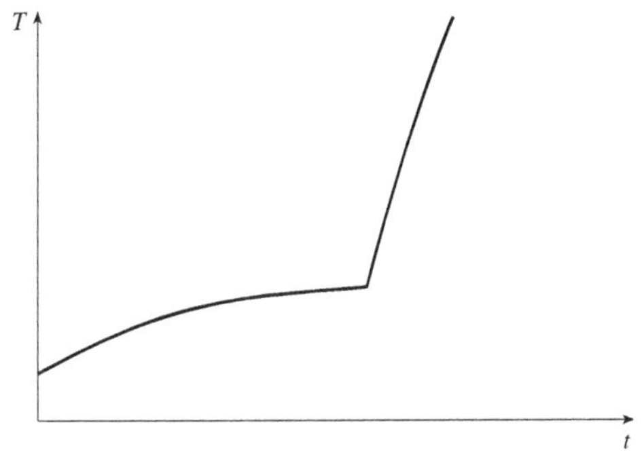

Graph 1

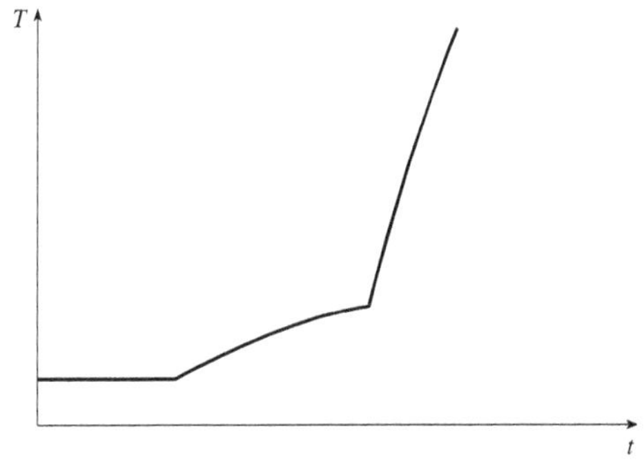

Graph 2

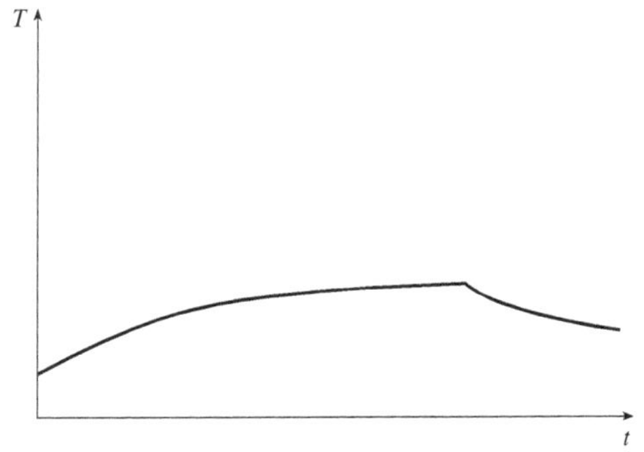

Graph 3

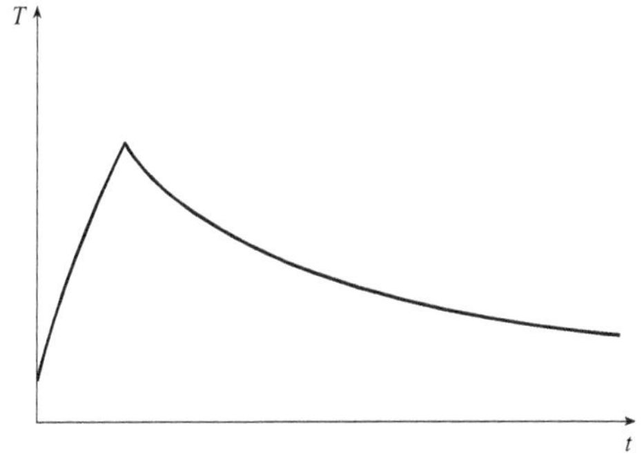

Graph 4

(a) I took my roast out of the freezer at noon, and left it on the counter to thaw. Then I cooked it in the oven when I got home.

(b) I took my roast out of the freezer this morning, and left it on the counter to thaw. Then I cooked it in the oven when I got home.

(c) I took my roast out of the freezer this morning, and left it on the counter to thaw. I forgot about it, and went out for Chinese food on my way home from work. I put it in the refrigerator when I finally got home.

GROUP WORK 2, SECTION 1.4

Functions in the Classroom

Which of the following relations are functions?

Domain	Function Values	Function
All the people in your classroom	Chairs	f (person) = his or her chair
All the people in your classroom	The set {blue, brown, green, hazel}	f (person) = his or her eye color
All the people in your classroom	Cities	f (person) = birthplace of their mom and dad
All the people in your classroom	$\mathbb{R}$, the real numbers	f (person) = number of molecules in their body
All the people in your classroom	Spleens	f (person) = his or her own spleen
All the people in your classroom	Pencils	f (person) = his or her pencil
All the people in the United States	Integers from 0–999999999	f (person) = his or her Social Security number
All the living people born in February	Days in February, 2007	f (person) = his or her birthday in February 2007
All the people in your classroom	Days of the year	f (person) = his or her birthday
All the people in your classroom	Cars	f (person) = his or her car
All the people in your classroom	$\mathbb{R}$, the real numbers	f (person) = how much cash he or she has
All the people in your college	Names	f (person) = his or her middle name
All the people in your classroom	People	f (person) = himself or herself
All the people in your classroom	People	f (person) = his or her algebra instructor

GROUP WORK 3, SECTION 1.4

Rational Functions

The functions below are sad and lonely because they have lost their graphs! Help them out by matching each function with its graph. One function's graph is not pictured here; when you are done matching, go ahead and sketch that function's graph.

1. $\dfrac{x^3 - x}{0.125}$

2. $\dfrac{x^2 - 1}{x^2 - 4}$

3. $\dfrac{x^2 - 4}{x^2 - 1}$

4. $\dfrac{x^2 - 1}{x + 0.5}$

5. $\dfrac{5x\left(x^2 - 1\right)}{x^2 + 1}$

6. $\dfrac{5x\left(x^2 - 1\right)}{-x^2 - 1}$

(a)

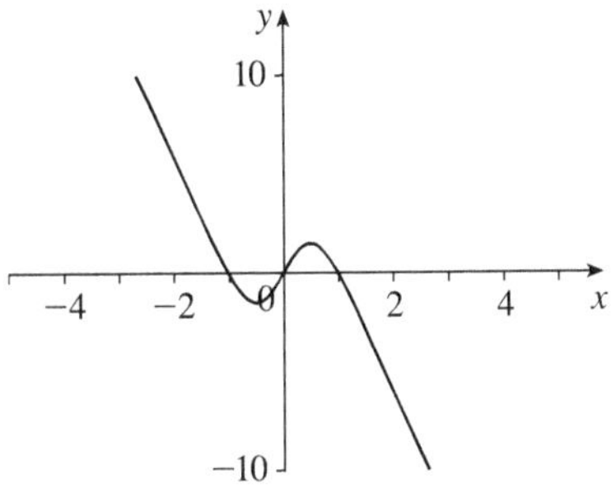

(b)

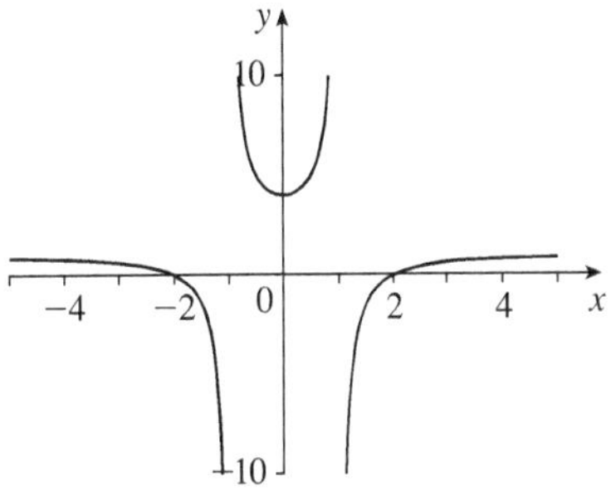

(c)

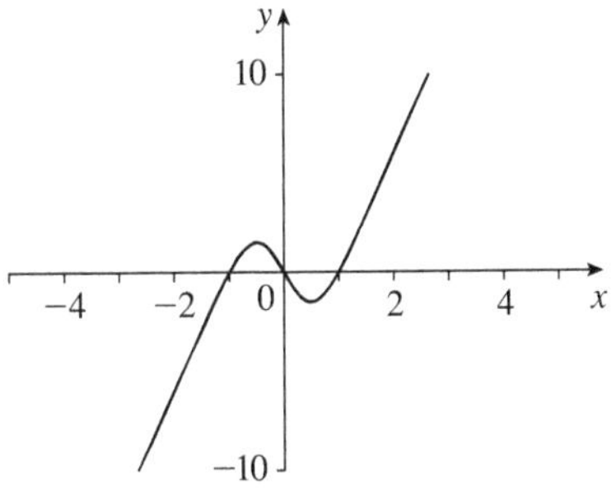

(d)

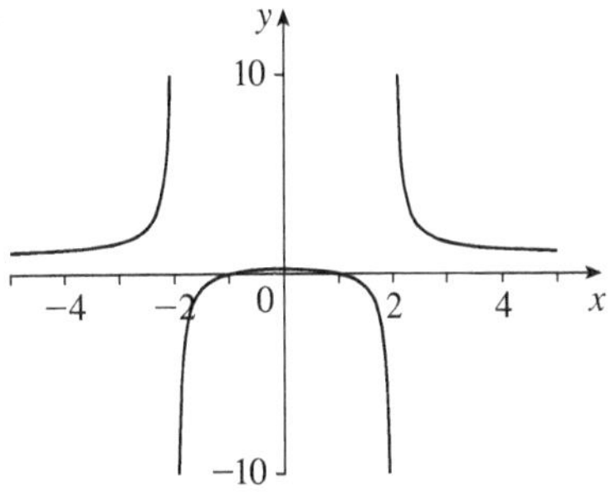

(e)

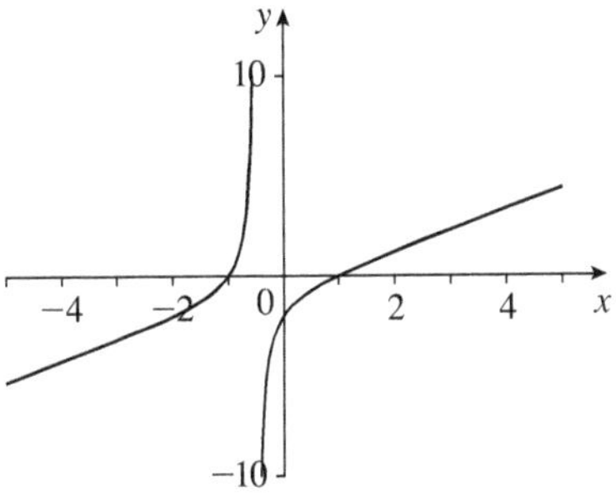

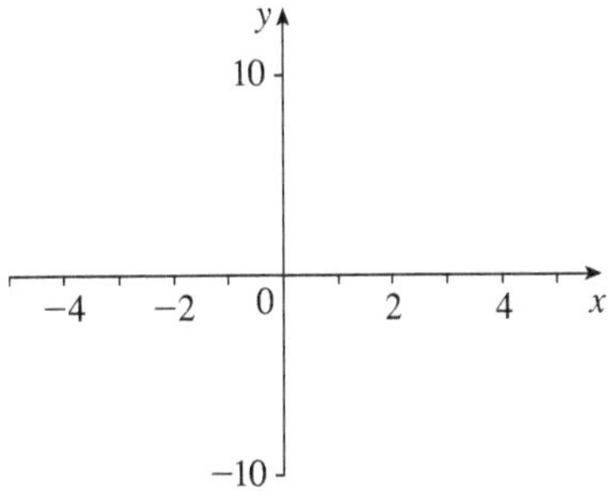

1.5 GETTING INFORMATION FROM THE GRAPH OF A FUNCTION

▼ Suggested Time and Emphasis

$\frac{1}{2}$–1 class. Essential material.

▼ Points to Stress

1. Gaining information about a function from its graph, including finding function values, domain and range.
2. Algebraic and geometric definitions of increasing and decreasing.
3. Finding local extrema of a function from its graph.

▼ Sample Questions

- **Text Question:** Draw a graph of a function with domain $[-10, 10]$ and range $[-2, 2]$. There should be at least one interval where the graph is increasing and at least one interval where the graph is decreasing.

  Answer: Answers will vary.

- **Drill Question:** If $f(x) = -x^2 + 9x + 2$, find the extreme value of f. Is it a maximum or a minimum?
  Answer: $f\left(\frac{9}{2}\right) = \frac{89}{4}$ is a maximum.

▼ In-Class Materials

- Explore domain and range with some graphs that have holes, such as the graphs of some of the functions in the previous section.

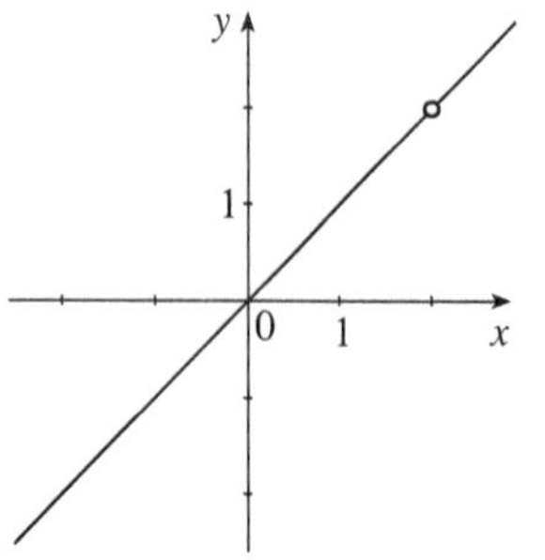

$f(x) = \dfrac{x(x-2)}{x-2}$

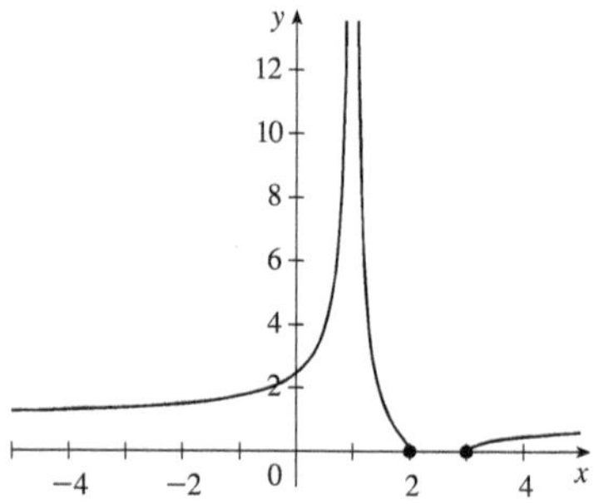

$g(x) = \sqrt{\dfrac{x^2 - 5x + 6}{x^2 - 2x + 1}}$

- Draw a graph of electrical power consumption in the classroom versus time on a typical weekday, pointing out important features throughout, and using the vocabulary of this section as much as possible.

- Notice that it is fairly easy to tell where some functions are increasing and decreasing by looking at their graphs. For example, the graph of $f(x) = x^4 - 8x^2$ makes things clear. Note that in this case, the intervals are not immediately apparent from looking at the formula. However, for many functions such as $g(x) = x^3 - 3x^2 + x + 1$, it is very difficult to find the exact intervals where the function is increasing/decreasing. In this example, the endpoints of the intervals will occur at precisely $x = 1 \pm \frac{1}{3}\sqrt{6}$.

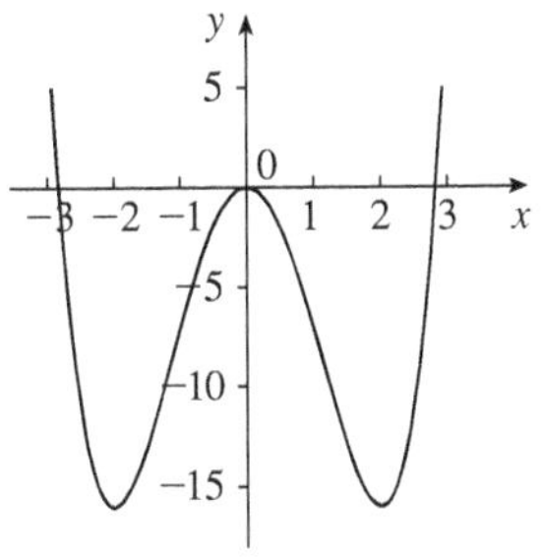

$f(x) = x^4 - 8x^2$

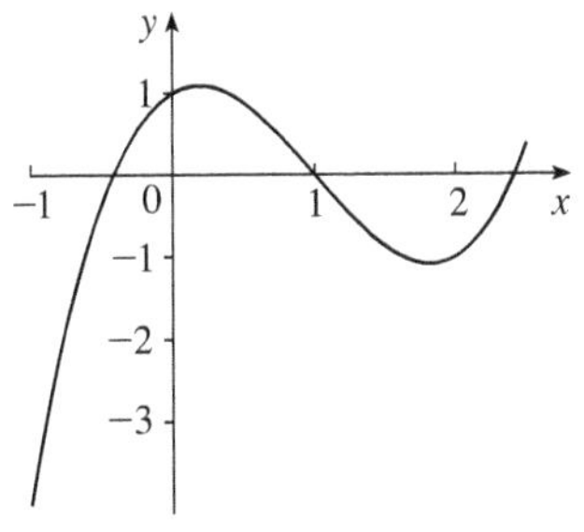

$g(x) = x^3 - 3x^2 + x + 1$

- Examine $f(x) = \begin{cases} x^2 & \text{if } x \text{ is rational} \\ 0 & \text{if } x \text{ is irrational} \end{cases}$ pointing out that it is neither increasing nor decreasing near $x = 0$. Stress that when dealing with new sorts of functions, it becomes important to know the precise mathematical definitions of such terms.

▼ Examples

- A function with two integer turning points and a flat spot:

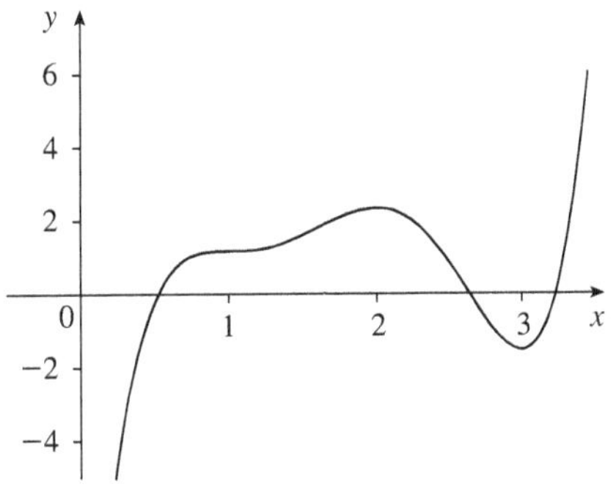

$\frac{1}{6}\left(12x^5 - 105x^4 + 340x^3 - 510x^2 + 360x - 90\right)$

- A function with several local extrema: $f(x) = x^4 + x^3 - 7x^2 - x + 6 = (x+3)(x+1)(x-1)(x-2)$

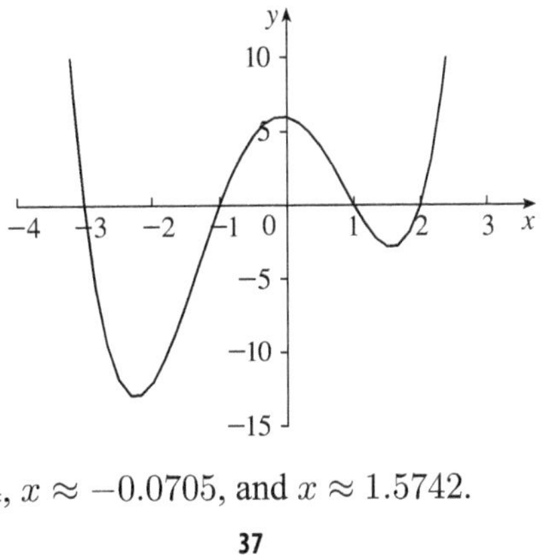

The extrema occur at $x \approx -2.254$, $x \approx -0.0705$, and $x \approx 1.5742$.

▼ Group Work 1: Calculator Exploration

This gives students a chance to graph things on their calculator and make conclusions. It will also serve as a warning that relying on calculator graphs without understanding the functions can lead one astray.

Notice that in calculus, when we say a function is increasing, we are saying it is increasing at every point on its domain. In this context, we are talking about increasing over an interval, which is slightly different. The curve $-1/x$, for example, is increasing at every point in its domain. Can we say it is decreasing over the interval $[-10, 1]$? No, because it is not defined in that interval. So the curve $-1/x$ is increasing over every interval for which it is defined.

Answers:

1. (c) **2.** (a) **3.** (a) (assuming positive intervals) **4.** (c) **5.** (c) **6.** (a) **7.** (b) **8.** (a) **9.** (c)

10. (c)

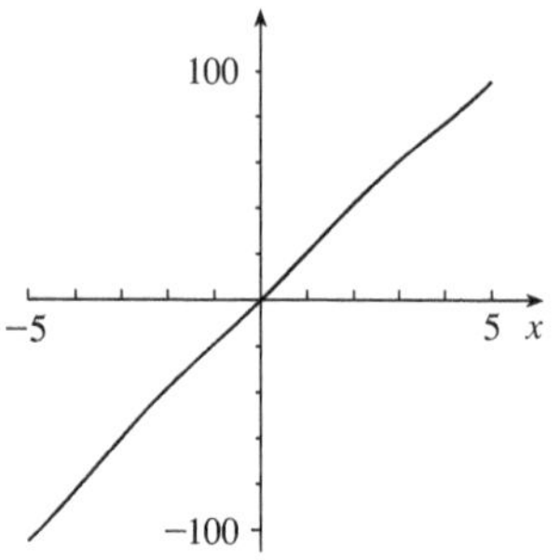

$f(x) = 20x + x \sin x$

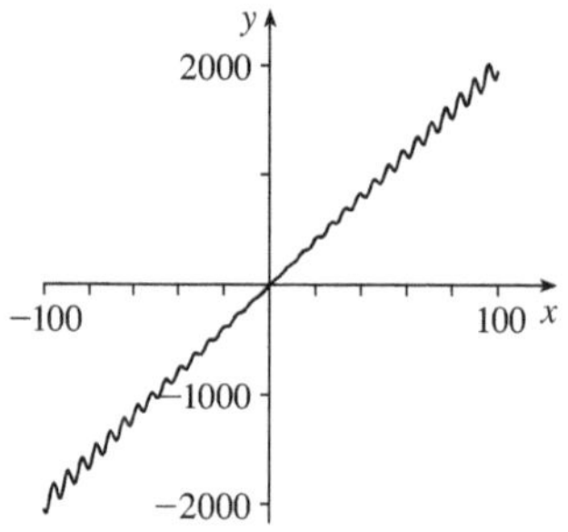

$f(x) = 20x + x \sin x$

▼ Group Work 2: The Little Dip

In this exercise students analyze a function with some subtle local extrema. After they have tried, reveal that there are two local maxima and two local minima.

After students have found the extrema, point out that if they take calculus, they will learn a relatively simple way to find the exact coordinates of the extrema.

Answers:

1.

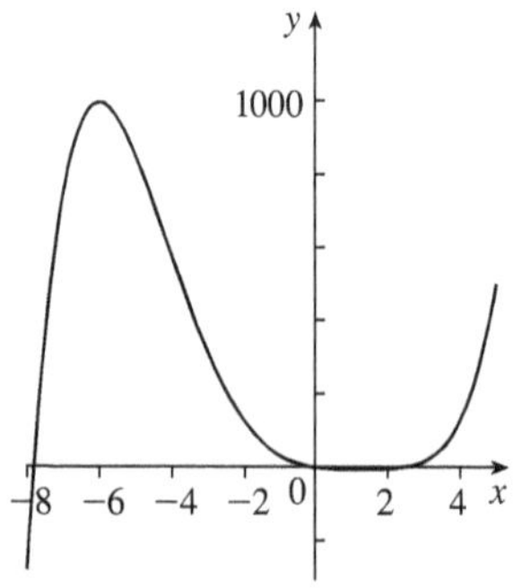

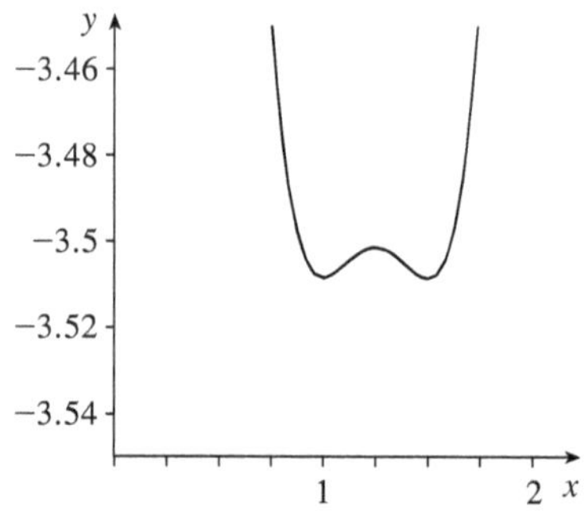

2. There are local maxima at $x = -6$ and $x = \frac{5}{4}$, and local minima at $x = 1$ and $x = \frac{3}{2}$.

▼ Homework Problems

Core Exercises: 3, 7, 8, 20, 28, 31, 37, 43, 49

Sample Assignment: 3, 4, 7, 8, 16, 20, 21, 23, 28, 31, 37, 43, 46, 49, 55

GROUP WORK 1, SECTION 1.5

Calculator Exploration

Graph the following curves on your calculator. For each curve specify which of the following applies.

(a) The graph of f is increasing over every interval (assuming the curve is defined everywhere in that interval).

(b) The graph of f is decreasing over every interval (assuming the curve is defined everywhere in that interval).

(c) The graph of f is increasing over some intervals and decreasing over others.

1. $f(x) = x^2$

2. $f(x) = x^3$

3. $f(x) = \sqrt{x}$

4. $f(x) = \sin x$

5. $f(x) = \cos x$

6. $f(x) = \tan x$

7. $f(x) = e^{-x}$

8. $f(x) = \ln x$

9. $f(x) = 5x^4 - 1.01^x$

10. $f(x) = 20x + x \sin x$

GROUP WORK 2, SECTION 1.5

The Little Dip

Consider $f(x) = \frac{1}{5}x^5 + \frac{9}{16}x^4 - \frac{143}{24}x^3 + \frac{207}{16}x^2 - \frac{45}{4}x$.

1. Draw a graph of f.

2. Estimate the x-values of all local extrema. Make sure your estimates are accurate to three decimal places.

1.6 TRANSFORMATIONS OF FUNCTIONS

▼ Suggested Time and Emphasis

1 class. Essential material.

▼ Points to Stress

1. Transforming a given function to a different one by shifting, stretching, and reflection.
2. Using the technique of reflection to better understand the concepts of even and odd functions.

▼ Sample Questions

- **Text Question:** What is the difference between a vertical stretch and a vertical shift?

  Answer: A vertical stretch extends the graph in the vertical direction, changing its shape. A vertical shift simply moves the graph in the vertical direction, preserving its shape.
- **Drill Question:** Given the graph of $f(x)$ below, sketch the graph of $\frac{1}{2}f(x) + 1$.

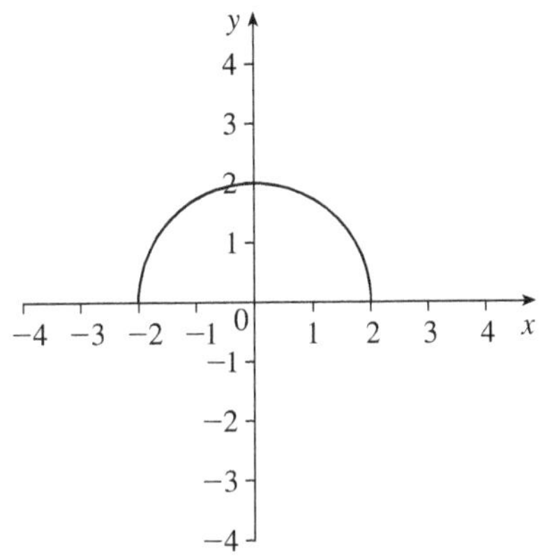

Answer:

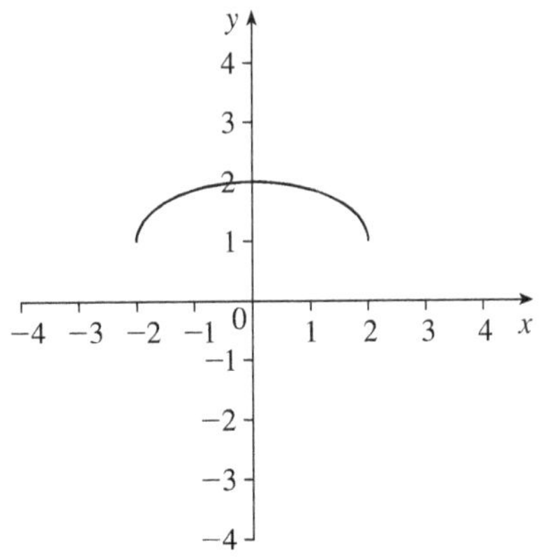

▼ In-Class Materials

- Students will often view this section as a process of memorizing eight similar formulas. Although it doesn't hurt to memorize how to shift, reflect, or stretch a graph, emphasize to students the importance of understanding what they are doing when they transform a graph. The group work "Discovering the Shift" (in Section 1.1) should help students understand and internalize. Tell students that if worse comes to worst, they can always plot a few points if they forget in which direction the graphs should move.

- Show the class a function they have not learned about yet, such as $f(x) = \sin x$. (If students know about sin, then show them arctan or e^{-x^2}—any function with which they are unfamiliar.) Point out that even though they don't know a lot about $\sin x$, once they've seen the graph, they can graph $\sin x + 3$, $\sin(x-1)$, $2\sin x$, $-\sin x$, etc.

- This is a good time to start discussing parameters. Ask your students to imagine a scientist who knows that a given function will be shaped like a stretched parabola, but has to do some more measurements to find out exactly what the stretching factor is. In other words, she can write $f(x) = -ax^2$, noting that she will have to figure out the a experimentally. The a is not a variable, it is a parameter. Similarly, if we are going to do a bunch of calculations with the function $f(x) = \sqrt[3]{x+2}$, and then do the same calculations with $\sqrt[3]{x+3}$, $\sqrt[3]{x-\pi}$, and $\sqrt[3]{x-\frac{2}{3}}$, it is faster and easier to do the set of calculations just once, with the function $g(x) = \sqrt[3]{x+h}$, and then fill in the different values for h at the end. Again, this letter h is called a parameter. Ask the class how, in the expression $f(t) = t + 3s$, they can tell which is the variable, and which is the parameter—the answer may encourage them to use careful notation.

▼ Examples

A distinctive-looking, asymmetric curve that can be stretched, shifted and reflected:

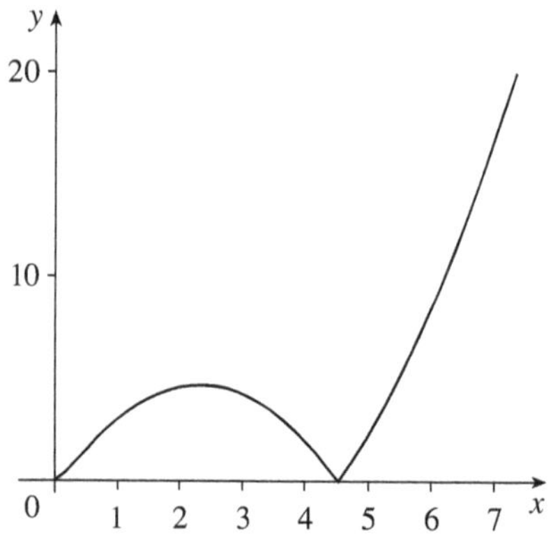

$f(x) = \left|x^2 - 5x + \sqrt{x}\right|$

▼ Group Work 1: Label Label Label, I Made It Out of Clay

Some of these transformations are not covered directly in the book. If the students are urged not to give up, and to use the process of elimination and testing individual points, they should be able to complete this activity.

Answers: 1. (d) **2.** (a) **3.** (f) **4.** (e) **5.** (i) **6.** (j) **7.** (b) **8.** (c) **9.** (g) **10.** (h)

▼ Group Work 2: Which is the Original?

The second problem has a subtle difficulty: the function is defined for all x, so some graphs show much more of the behavior of $f(x)$ than others do.

Answers: 1. $2f(x+2)$, $2f(x)$, $f(2x)$, $f(x+2)$, $f(x)$ **2.** $2f(x)$, $f(x)$, $f(x+2)$, $f(2x)$, $2f(x+2)$

▼ Homework Problems

Core Exercises: 4, 7, 11, 15, 20, 38, 44, 45, 48, 62, 77

Sample Assignment: 4, 6, 7, 11, 14, 15, 20, 38, 44, 45, 48, 51, 53, 58, 62, 77, 80, 83, 89

GROUP WORK 1, SECTION 1.6
Label Label Label, I Made it Out of Clay

This is a graph of the function $f(x)$:

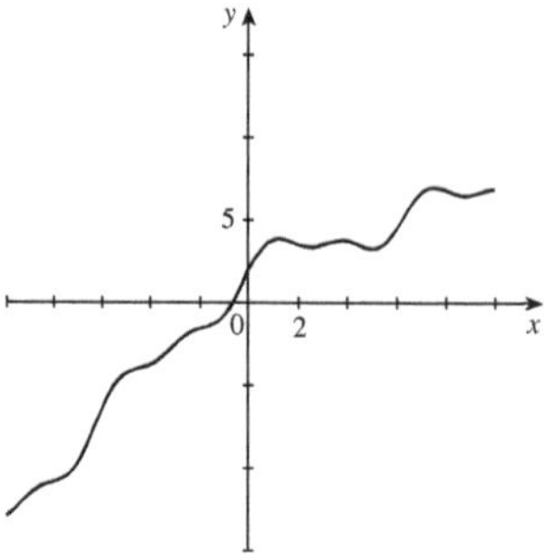

Give each graph below the correct label from the following:

(a) $f(x+3)$ (b) $f(x-3)$ (c) $f(2x)$ (d) $2f(x)$ (e) $|f(x)|$

(f) $f(|x|)$ (g) $2f(x)-1$ (h) $f(2x)+2$ (i) $f(x)-x$ (j) $1/f(x)$

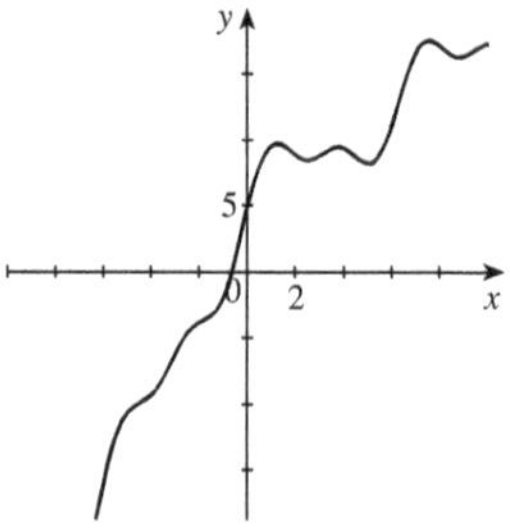

Graph 1

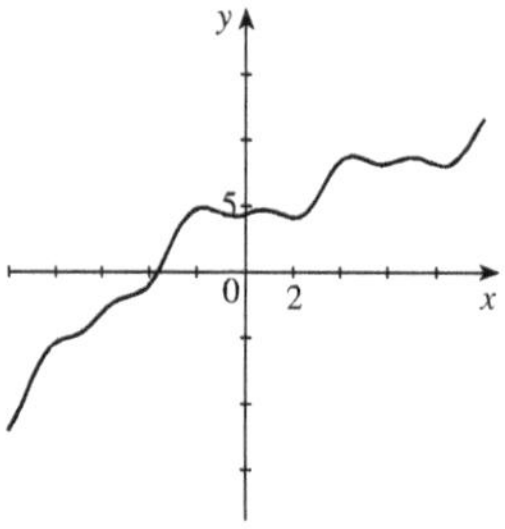

Graph 2

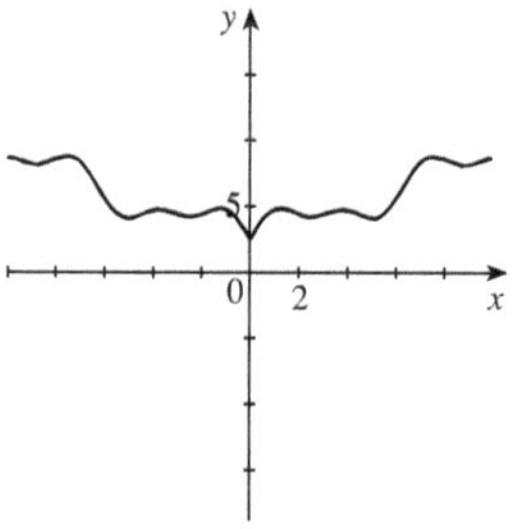

Graph 3

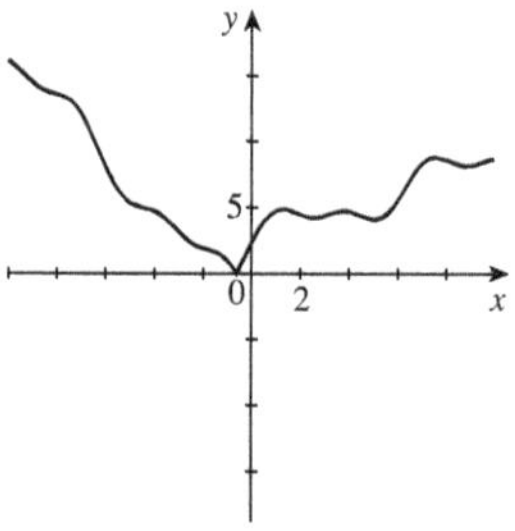

Graph 4

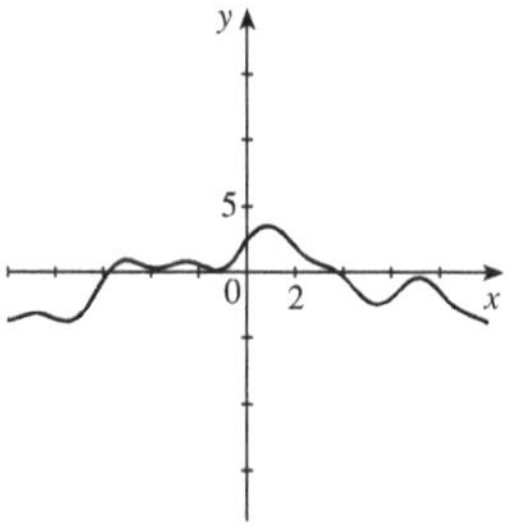

Graph 5

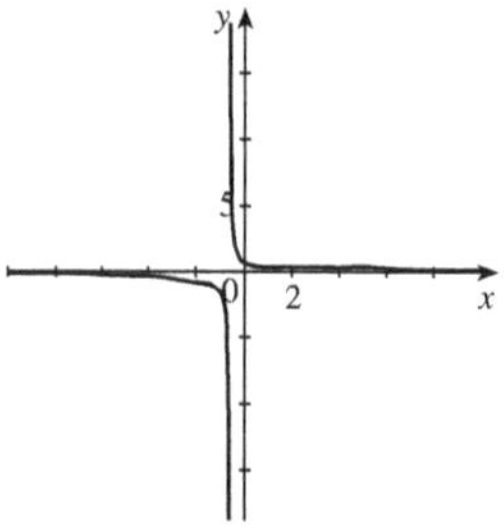

Graph 6

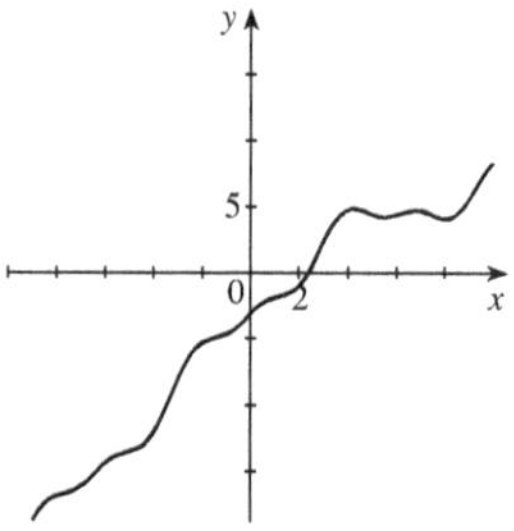

Graph 7

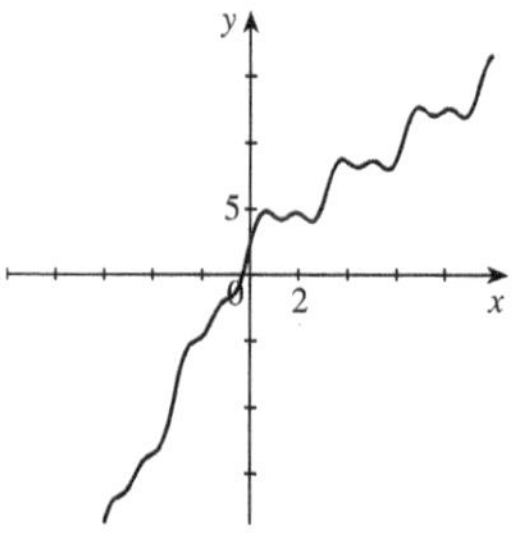

Graph 8

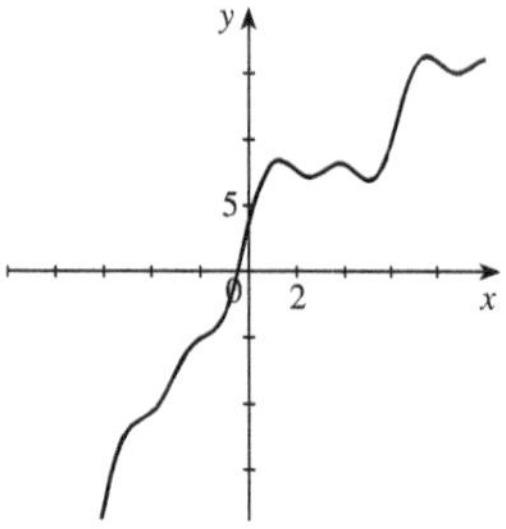

Graph 9

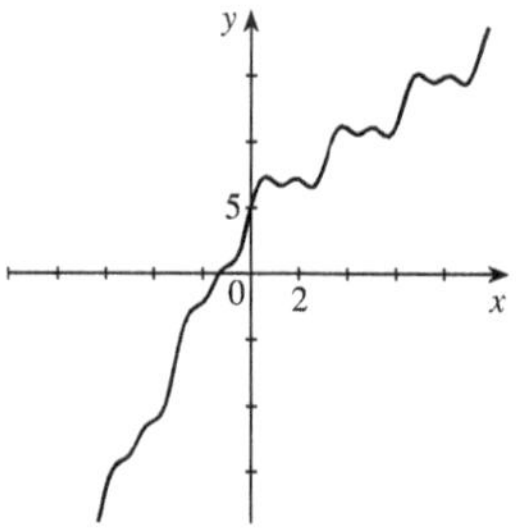

Graph 10

GROUP WORK 2, SECTION 1.6

Which is the Original?

Below are five graphs. One is the graph of a function $f(x)$ and the others include the graphs of $2f(x)$, $f(2x)$, $f(x+2)$, and $2f(x+2)$. Determine which is the graph of $f(x)$ and match the other functions with their graphs.

1.

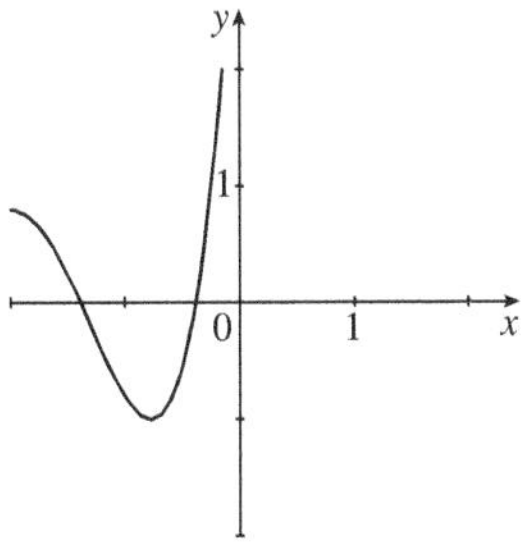

Graph 1

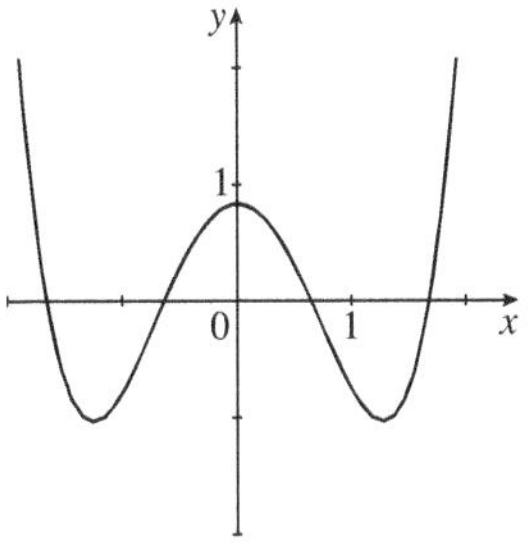

Graph 2

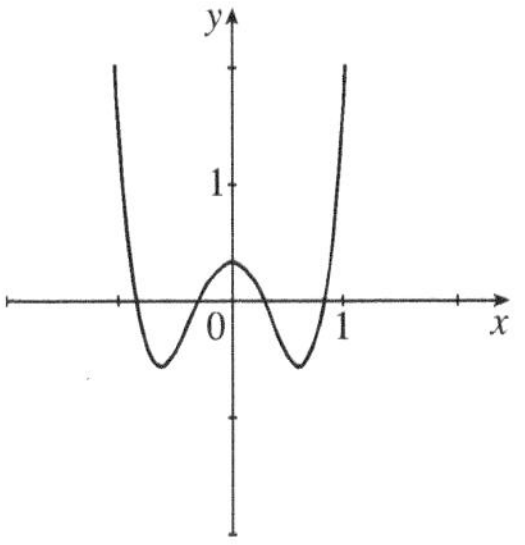

Graph 3

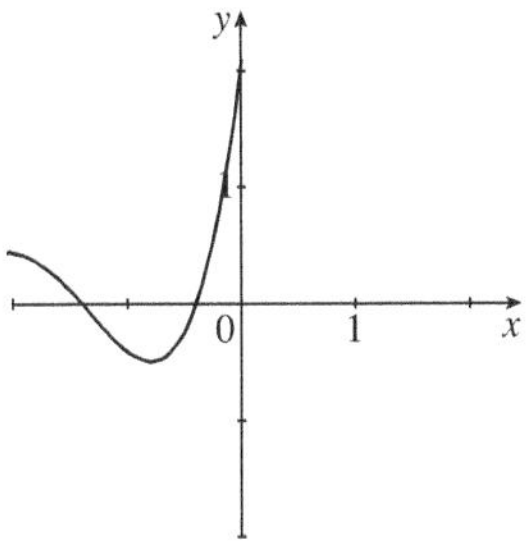

Graph 4

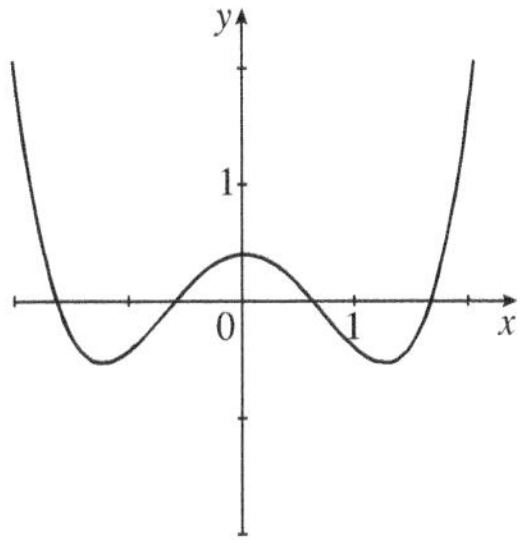

Graph 5

2.

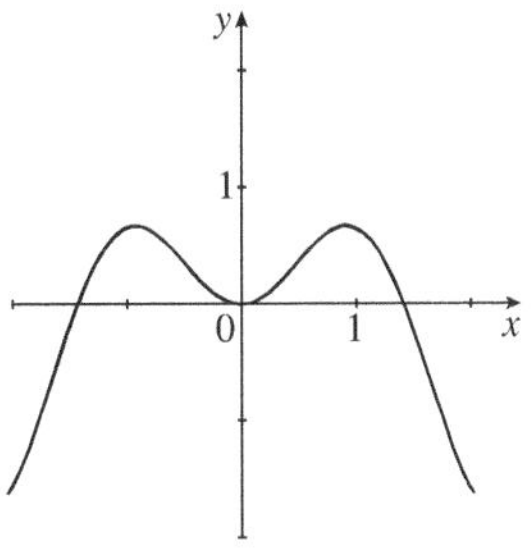

Graph 1

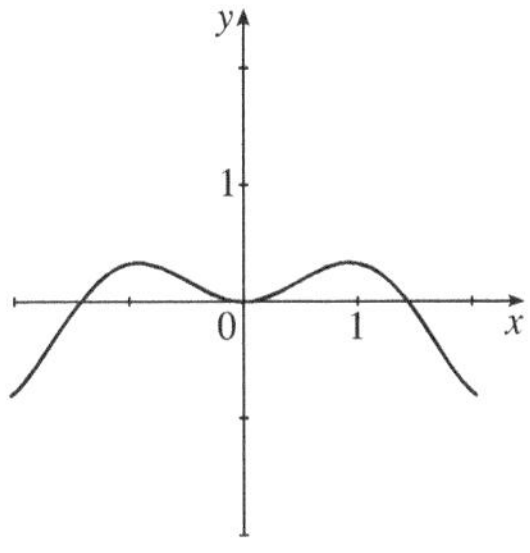

Graph 2

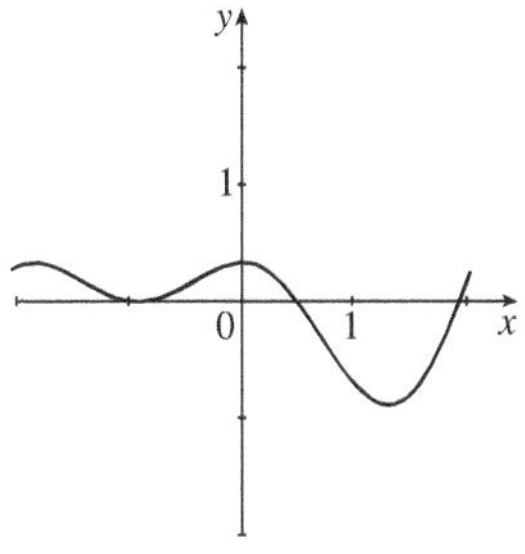

Graph 3

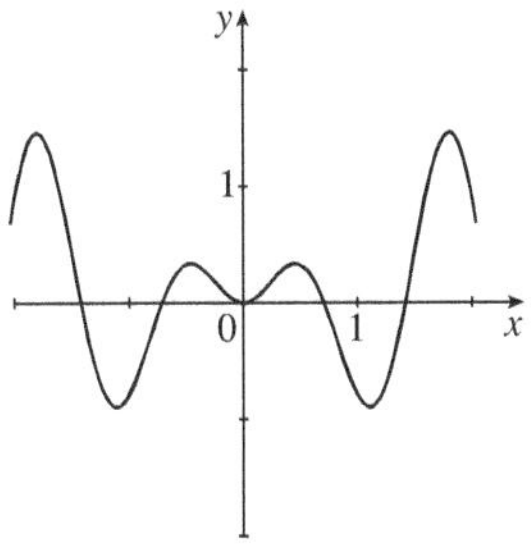

Graph 4

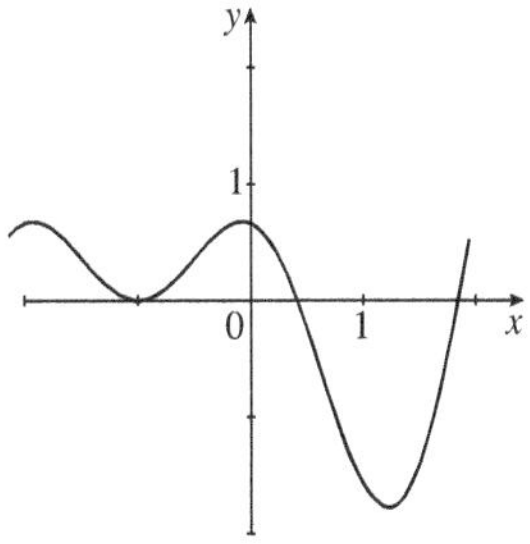

Graph 5

1.7 COMBINING FUNCTIONS

▼ Suggested Time and Emphasis

$\frac{1}{2}$–1 class. Essential material.

▼ Points to Stress

1. Addition, subtraction, multiplication, and division of functions.

2. Composition of functions.

3. Finding the domain of a function based on analysis of the domain of its components.

▼ Sample Questions

- **Text Question:** The text describes addition, multiplication, division, and composition of functions. Which of these operations is represented by the following diagram?

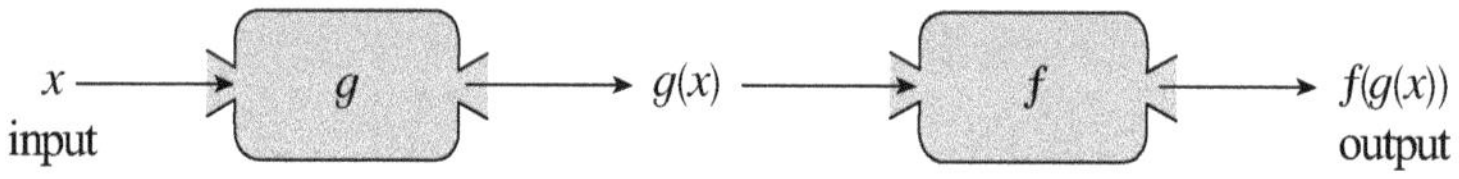

Answer: Composition

- **Drill Question:** Let $f(x) = 4x$ and $g(x) = x^3 + x$.

(a) Compute $(f \circ g)(x)$.

(b) Compute $(g \circ f)(x)$.

Answer: **(a)** $4(x^3 + x) = 4x^3 + 4x$ **(b)** $(4x)^3 + 4x = 64x^3 + 4x$

▼ In-Class Materials

- Do the following problem with the class:

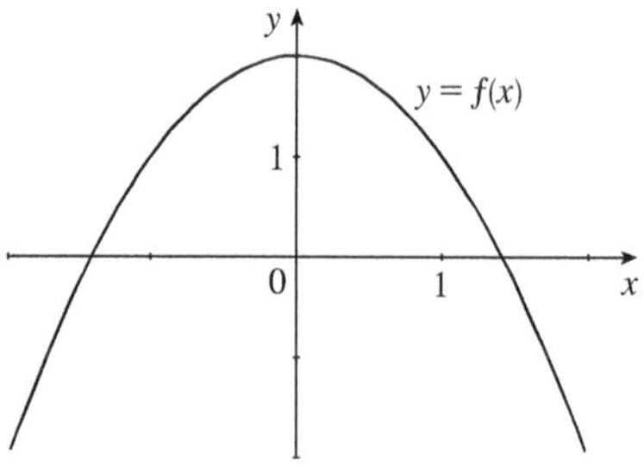

From the graph of $y = f(x) = -x^2 + 2$ shown above, compute $f \circ f$ at $x = -1$, 0, and 1. First do it graphically, then algebraically.

- Show the tie between algebraic addition of functions and graphical addition. For example, let $f(x) = 1 - x^2$ and $g(x) = x^2 + \frac{1}{2}x - 1$. First add the functions graphically, as shown below, and then show how this result can be obtained algebraically: $(1 - x^2) + (x^2 + \frac{1}{2}x - 1) = \frac{1}{2}x$.

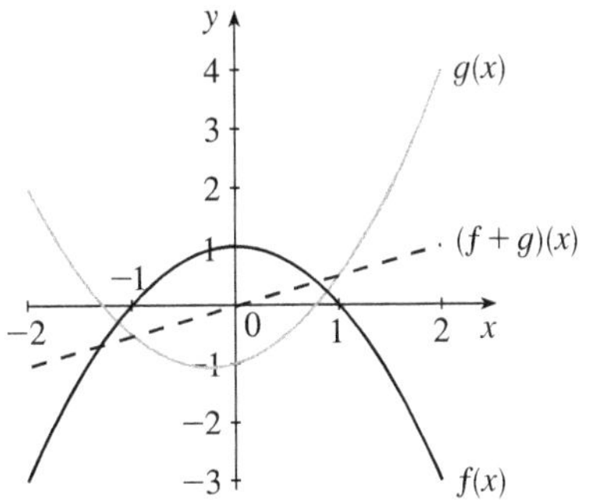

- Point out that it is important to keep track of domains, especially when doing algebraic simplification. For example, if $f(x) = x + \sqrt{x}$ and $g(x) = 3x^2 + \sqrt{x}$, even though $(f - g)(x) = x - 3x^2$, its domain is not $\mathbb{R}$ but $\{x \mid x \geq 0\}$.

- Function maps are a nice way to explain composition of functions. To demonstrate $g \circ f(x)$, draw three number lines labeled x, $f(x)$, and $g(x)$, and then indicate how each number x goes to $f(x)$ which then goes to $g(f(x))$. For example, if $f(x) = \sqrt{x}$ and $g(x) = 2x - 1$, the diagram looks like this:

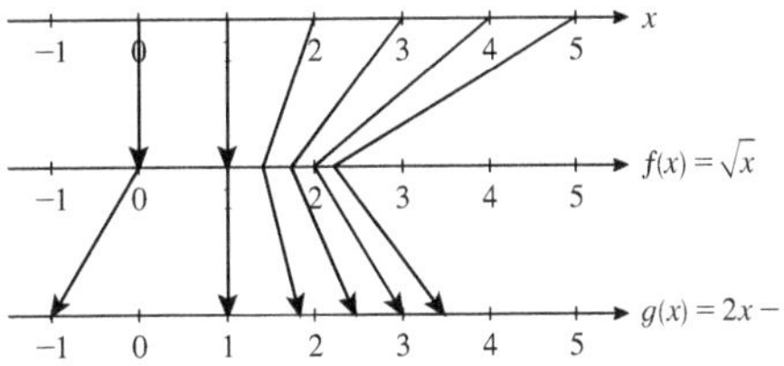

- After doing a few basic examples of composition, it is possible to foreshadow the idea of inverses, which will be covered in the next section. Let $f(x) = 2x^3 + 3$ and $g(x) = x^2 - x$. Compute $f \circ g$ and $g \circ f$ for your students. Then ask them to come up with a function $h(x)$ with the property that $(f \circ h)(x) = x$. They may not be used to the idea of coming up with examples for themselves, so the main hints they will need might be "don't give up," "when in doubt, just try something and see what happens," and "I'm not expecting you to get it in fifteen seconds." If the class is really stuck, have them try $f(x) = 2x^3$ to get a feel for how the game is played. Once they have determined that $h(x) = \sqrt[3]{\dfrac{x-3}{2}}$, have them compute $(h \circ f)(x)$ and have them conjecture whether, in general, if $(f \circ g)(x) = x$ then $(g \circ f)(x)$ must also equal x.

▼ Examples

Combined functions with graphs: Let $f(x) = x^2 - 3x + 2$ and $g(x) = -\sqrt{x}$.

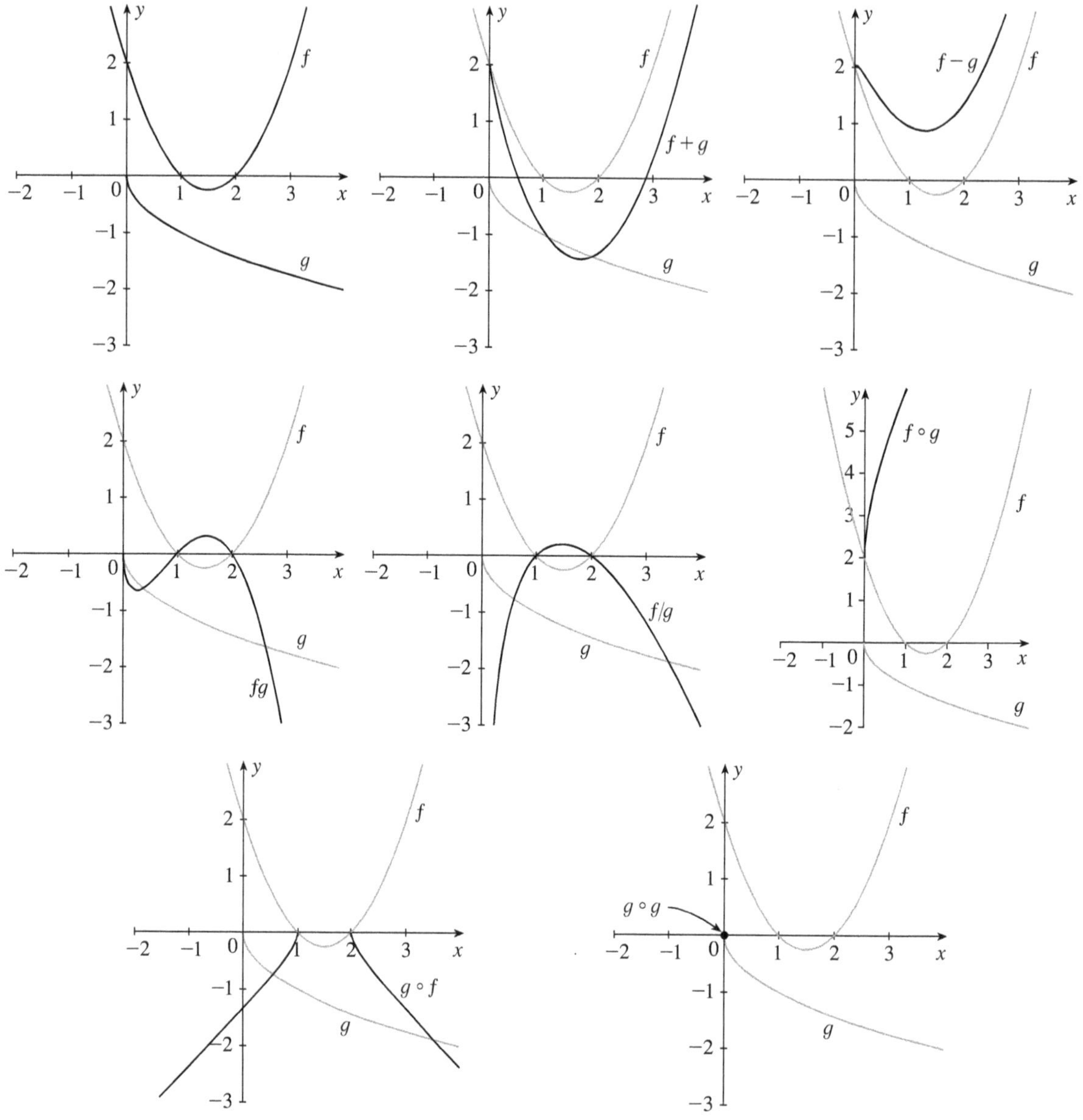

▼ Group Work 1: Transformation of Plane Figures

This tries to remove the composition idea from the numerical context, and introduces the notion of symmetry groups. It is a longer activity than it seems, and can lead to an interesting class discussion of this topic.

Answers:

1. 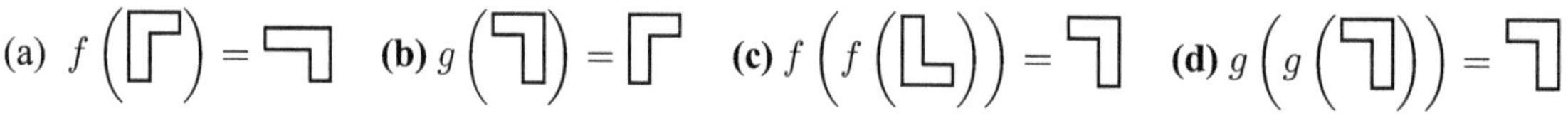

2. This is false. For example, $(f \circ g)$  = , but $(g \circ f)$ 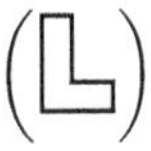 = .

3. It is true: reversing something thrice in a mirror gives the same result as reversing it once.

4. It rotates the shape 270° clockwise or, equivalently, 90° counterclockwise.

▼ Group Work 2: Odds and Evens

This is an extension of Exercise 70 in the text. Students may find the third problem difficult to start. You may want to give selected table entries on the board first, before handing the activity out, to make sure students understand what they are trying to do.

Answers:

1.

a	b	$a+b$
even	even	even
odd	even	odd
even	odd	odd
odd	odd	even

2.

a	b	$a \cdot b$
even	even	even
odd	even	even
even	odd	even
odd	odd	odd

3.

f	g	$f+g$	fg	$f \circ g$	$g \circ f$
even	even	even	even	even	even
even	odd	neither	odd	even	even
odd	even	neither	odd	even	even
odd	odd	odd	even	odd	odd
neither	neither	unknown	unknown	unknown	unknown

▼ Group Work 3: It's More Fun to Compute

Each group gets one copy of the graph. During each round, one representative from each group stands, and one of the questions below is asked. The representatives write their answer down, and all display their answers at the same time. Each representative has the choice of consulting with their group or not. A correct solo answer is worth two points, and a correct answer after a consult is worth one point.

1. $(f \circ g)(5)$
2. $(g \circ f)(5)$
3. $(f \circ g)(0)$
4. $(f \circ f)(5)$
5. $(g \circ g)(5)$
6. $(g \circ g)(-3)$
7. $(g \circ g)(-1)$
8. $(f \circ g)(1)$
9. $(g \circ f)(1)$
10. $(f \circ f \circ g)(4)$
11. $(g \circ f \circ f)(4)$
12. $(f \circ g \circ f)(4)$

Answers: **1.** 0 **2.** 0 **3.** 1 **4.** 5 **5.** 1 **6.** 1 **7.** 1 **8.** 0 **9.** 2 **10.** 1 **11.** 1 **12.** 1

▼ Homework Problems

Core Exercises: 1, 6, 7, 16, 21, 25, 29, 38, 46, 54

Sample Assignment: 1, 6, 7, 10, 11, 16, 21, 25, 26, 29, 38, 44, 46, 54, 58, 66, 68

GROUP WORK 1, SECTION 1.7

Transformation of Plane Figures

So far, when we have been talking about functions, we have been assuming that their domains and ranges have been sets of numbers. This is not necessarily the case. For example, look at this figure:

Let's let our domain be all the different ways we can move this figure around, including flipping it over:

$D =$ 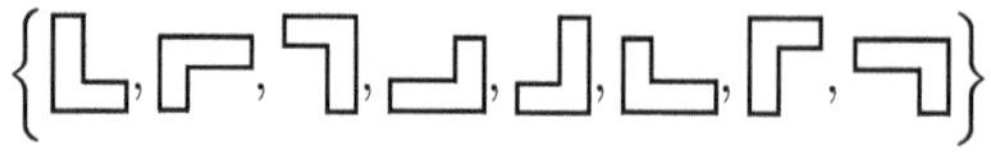

Now let f be the function that rotates the shape 90° clockwise: f  $=$ . Let g be the function that flips the shape over a vertical line drawn through the center: g  $=$ 

1. Find the following:

(a) f 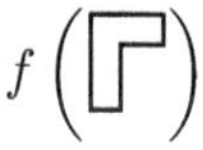   **(b)** g    **(c)** $f(f($  $))$ **(d)** $g(g($  $))$

2. Is it true that $f \circ g = g \circ f$? Why or why not?

3. Is it true that $g \circ g \circ g = g$? Why or why not?

4. Write, in words, what the function $f \circ f \circ f$ does to a shape.

GROUP WORK 2, SECTION 1.7

Odds and Evens

1. Let a be an odd number, and b be an even number. Fill in the following table (the first row is done for you).

a	b	$a+b$
even	even	even
odd	even	
even	odd	
odd	odd	

2. We can also multiply numbers together. Fill in the corresponding multiplication table:

a	b	$a \cdot b$
even	even	
odd	even	
even	odd	
odd	odd	

3. Now we let f and g be (nonzero) functions, not numbers. We are going to think about what happens when we combine these functions. When you fill in the table, you can write "unknown" if the result can be odd *or* even, depending on the functions. You can solve this problem by drawing some pictures, or by using the definition of odd and even functions.

f	g	$f+g$	fg	$f \circ g$	$g \circ f$
even	even				
even	odd				
odd	even				
odd	odd				
neither	neither				

GROUP WORK 3, SECTION 1.7

It's More Fun to Compute

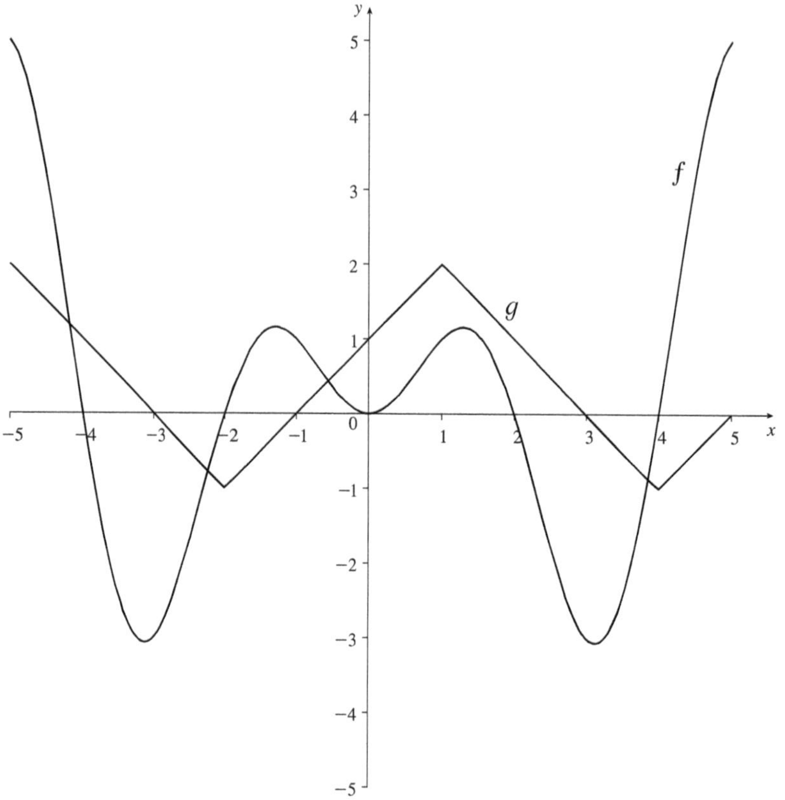

1.8 ONE-TO-ONE FUNCTIONS AND THEIR INVERSES

▼ Suggested Time and Emphasis

1–2 classes. Essential material.

▼ Points to Stress

1. One-to-one functions: their definition and the Horizontal Line Test.
2. Algebraic and geometric properties of inverse functions.
3. Finding inverse functions.

▼ Sample Questions

- **Text Question:** The function f is graphed below. Sketch f^{-1}, the inverse function of f.

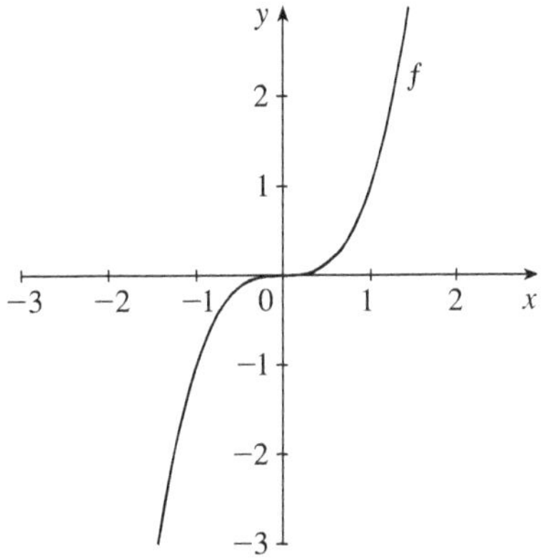

Answer:

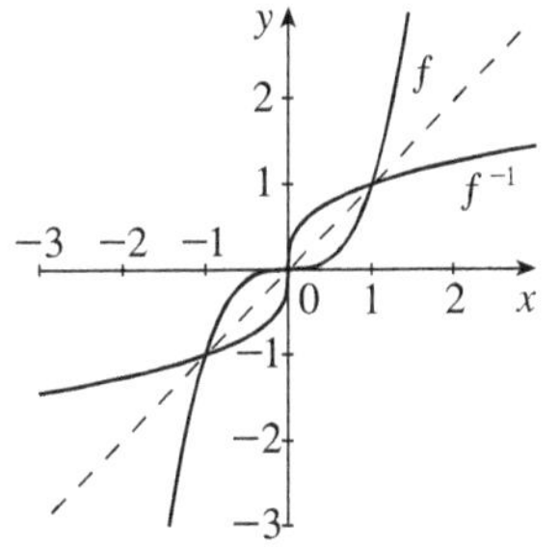

- **Drill Question:** If $f(-2) = 4$, $f(-1) = 3$, $f(0) = 2$, $f(1) = 1$ and $f(2) = 3$, what is $f^{-1}(2)$?

  Answer: 0

▼ In-Class Materials

- Make sure students understand the notation: f^{-1} is not the same thing as $\frac{1}{f}$.
- Starting with $f(x) = \sqrt[3]{x-4}$, compute $f^{-1}(-2)$ and $f^{-1}(0)$. Then use algebra to find a formula for $f^{-1}(x)$. Have the class try to repeat the process with $g(x) = x^3 + x - 2$. Note that facts such as $g^{-1}(-2) = 0$, $g^{-1}(0) = 1$, and and $g^{-1}(8) = 2$ can be found by looking at a table of values for $g(x)$ but that algebra fails to give us a general formula for $g^{-1}(x)$. Finally, draw graphs of f, f^{-1}, g, and g^{-1}.

- Pose the question: If f is always increasing, is f^{-1} always increasing? Give students time to try prove their answer.

  Answer: This is true. Proofs may involve diagrams and reflections about $y = x$, or you may try to get them to be more rigorous. This is an excellent opportunity to discuss concavity, noting that if f is concave up and increasing, then f^{-1} is concave down and increasing.

- Point out that the idea of "reversing input and output" permeates the idea of inverse functions, in all four representations of "function". When finding inverse functions algebraically, we explicitly reverse x and y. When drawing the inverse function of a graph, by reflecting across the line $y = x$ we are reversing the y- and x-axes. If $c(x)$ is the cost (in dollars) to make x fruit roll-ups, then $c^{-1}(x)$ is the number of fruit roll-ups that could be made for x dollars—again reversing the input and the output. Finally, show the class how to find the inverse of a function given a numeric data table, and note that again the inputs and outputs are reversed.

x	$f(x)$
1	3
2	4.2
3	5.7
4	8

x	$f^{-1}(x)$
3	1
4.2	2
5.7	3
8	4

- Make sure to discuss units carefully: when comparing $y = f(x)$ to $y = f^{-1}(x)$, the units of y and x trade places.

▼ Examples

- The graph of a complicated function and its inverse:

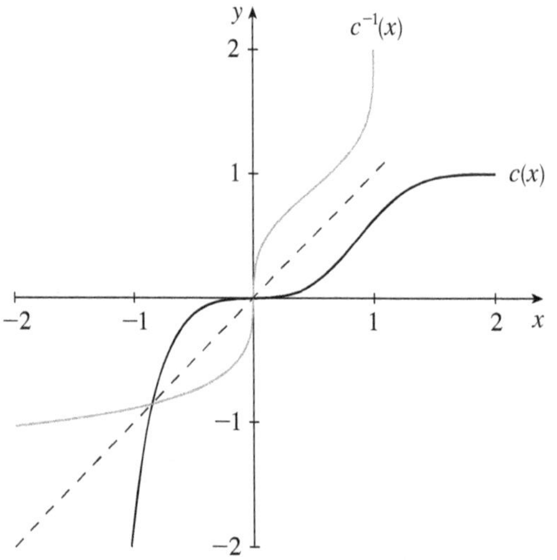

▼ Group Work 1: Inverse Functions: Domains and Ranges

While discussing the domains and ranges of inverse functions, this exercise foreshadows later excursions into the maximum and minimum values of functions.

If a group finishes early, ask them this question:

"Now consider the graph of $f(x) = \sqrt{2x-3}+2$. What are the domain and range of $f(x)$? Try to figure out the domain and range of $f^{-1}(x)$ by looking at the graph of f. In general, what information do you need to be able to compute the domain and range of $f^{-1}(x)$ from the graph of a function f?"

Answers:

1. It is one-to-one, because the problem says it climbs steadily.
2. a^{-1} is the time in minutes at which the plane achieves a given altitude.
3. Reverse the data columns in the given table to get the table for the inverse function. The domain and range of a are $0 \le t \le 30$ and $0 \le a \le 29{,}000$, so the domain and range of a^{-1} are $0 \le x \le 29{,}000$ and $0 \le a^{-1} \le 30$.
4. You can expect to turn on your computer after about 8.5 minutes.
5. a is no longer 1-1, because heights are now achieved more than once.

Bonus The domain of f^{-1} is the set of all y-values on the graph of f, and the range of f^{-1} is the set of all x-values on the graph of f.

▼ Group Work 2: The Column of Liquid

If students need a hint, you can mention that the liquid used in the drugstore is mercury.

Answers:

1. The liquid is 1 cm high when the temperature is 32° F.
2. The liquid is 10 cm high when the temperature is 212° F
3. The inverse function takes a height in cm, and gives the temperature. So it is a device for measuring temperature.
4. A thermometer

▼ Group Work 3: Functions in the Classroom Revisited

This activity starts the same as "Functions in the Classroom" from Section 1.4.2 At this point, students have learned about one-to-one functions, and they are able to explore this activity in more depth.

Answers:

Chairs: Function, one-to-one, bijection (if all chairs are occupied). If one-to-one, the inverse assigns a chair to a person.

Eye color: Function, not one-to-one

Mom & Dad's birthplace: Not a function; mom and dad could have been born in different places

Molecules: Function, one-to-one (with nearly 100% probability); inverse assigns a number of molecules to the appropriate student.

Spleens: Function, one-to-one, bijection. Inverse assigns each spleen to its owner.

Pencils: Not a function; some people may have more than one or (horrors!) none.

Social Security Number: Function, one-to-one; inverse assigns each number to its owner.

February birthday: Not a function; not defined for someone born on February 29.

Birthday: Function, perhaps one-to-one. If one-to-one, the inverse assigns a day to a person.

Cars: Not a function; some have none, some have more than one.

Cash: Function, perhaps one-to-one. If one-to-one, the inverse assigns an amount of money to a person.

Middle names: Not a function; some have none, some have more than one.

Identity: Function, one-to-one, bijection. Inverse is the same as the function.

Algebra instructor: Function, not one-to-one.

▼ Homework Problems

Core Exercises: 4, 5, 13, 17, 22, 28, 31, 37, 44, 54, 76

Sample Assignment: 4, 5, 10, 13, 17, 22, 28, 31, 34, 36, 37, 44, 45, 54, 60, 63, 75, 76, 85

GROUP WORK 1, SECTION 1.8

Inverse Functions: Domains and Ranges

Let $a(t)$ be the altitude in feet of a plane that climbs steadily from takeoff until it reaches its cruising altitude after 30 minutes. We don't have a formula for a, but extensive research has given us the following table of values:

t	$a(t)$
0.1	50
0.5	150
1	500
3	2000
7	8000
10	12,000
20	21,000
25	27,000
30	29,000

1. Is $a(t)$ a one-to-one function? How do you know?

2. What does the function a^{-1} measure in real terms? Your answer should be descriptive, similar to the way $a(t)$ was described above.

3. We are interested in computing values of a^{-1}. Fill in the following table for as many values of x as you can. What quantity does x represent?

x	$a^{-1}(x)$

What are the domain and range of a? What are the domain and range of a^{-1}?

4. You are allowed to turn on electronic equipment after the plane has reached 10,000 feet. Approximately when can you expect to turn on your laptop computer after taking off?

5. Suppose we consider $a(t)$ from the time of takeoff to the time of touchdown. Is $a(t)$ still one-to-one?

GROUP WORK 2, SECTION 1.8

The Column of Liquid

It is a fact that if you take a tube and fill it partway with liquid, the liquid will rise and fall based on the temperature. Assume that we have a tube of liquid, and we have a function $h(T)$, where h is the height of the liquid in cm at temperature T in °F.

1. It is true that $h(32) = 1$. What does that mean in physical terms?

2. It is true that $h(212) = 10$. What does that mean in physical terms?

3. Describe the inverse function h^{-1}. What are its inputs? What are its outputs? What does it measure?

4. There is a device, currently available at your local drugstore, that measures the function h^{-1}. What is the name of this device?

GROUP WORK 3, SECTION 1.8

Functions in the Classroom Revisited

Which of the following are functions? Of the ones that are functions, which are one-to-one functions? Describe what the inverses tell you.

Domain	Function Values	Function
All the people in your classroom	Chairs	f (person) = his or her chair
All the people in your classroom	The set {blue, brown, green, hazel}	f (person) = his or her eye color
All the people in your classroom	Cities	f (person) = birthplace of their mom and dad
All the people in your classroom	$\mathbb{R}$, the real numbers	f (person) = number of molecules in their body
All the people in your classroom	Spleens	f (person) = his or her own spleen
All the people in your classroom	Pencils	f (person) = his or her pencil
All the people in the United States	Integers from 0–999999999	f (person) = his or her Social Security number
All the living people born in February	Days in February, 2007	f (person) = his or her birthday in February 2007
All the people in your classroom	Days of the year	f (person) = his or her birthday
All the people in your classroom	Cars	f (person) = his or her car
All the people in your classroom	$\mathbb{R}$, the real numbers	f (person) = how much cash he or she has
All the people in your college	Names	f (person) = his or her middle name
All the people in your classroom	People	f (person) = himself or herself
All the people in your classroom	People	f (person) = his or her algebra instructor

2 TRIGONOMETRIC FUNCTIONS: UNIT CIRCLE APPROACH

2.1 THE UNIT CIRCLE

▼ Suggested Time and Emphasis

$\frac{1}{2}$–1 class. Essential material.

▼ Points to Stress

1. Finding the reference number for a given real number.
2. Using reference numbers to find terminal points.

▼ Sample Questions

- **Text Question:** Is it possible for two different real numbers to have the same reference number?

  Answer: Yes

- **Drill Question:** Find the reference number $\bar{t}$ for $t = \frac{17\pi}{6}$.

  Answer: $\bar{t} = \frac{\pi}{6}$

▼ In-Class Materials

- Look at the time, and then ask the class what time it will be 8 hours from now. Then ask about 12 hours from now, 13 hours from now, 25 hours from now, and 50 hours from now. Then ask what time it was 12 hours ago, 11 hours ago, and 13 hours ago. Draw the analogy of the type of arithmetic you are doing when computing future and past times to that of finding reference angles, only instead of freely adding and subtracting 12, we are freely adding and subtracting 2π.
- Figure 8 is crucial, because the way we find $\bar{t}$ depends upon the quadrant we are in. Perhaps find the reference numbers for $t = \frac{\pi}{3}, \frac{2\pi}{3}, \frac{3\pi}{3} = \pi, \frac{4\pi}{3}, \frac{5\pi}{3}$, and $\frac{6\pi}{3} = 2\pi$ all in a row to illustrate this principle.
- The text uses integer multiples of $\frac{\pi}{6}$ and $\frac{\pi}{4}$ for most of its examples of finding reference numbers. After the students understand these examples, try having them work with different rational multiples of π, such as $-\frac{121}{35}\pi$ or $\frac{10{,}001}{5}\pi$. Transition them to numbers whose terminal points lie in quadrants that are easy to figure out, such as $t = 3$, 6, or 1.5. Finally, have them try to find reference numbers of arbitrary real numbers such as 22 or -13.
- Point out that finding terminal points is much more difficult than finding reference numbers. If the reference number is not given in Table 1, they don't have a good method for finding the terminal point. Have them try to estimate a few by measurement. For example, have the students draw a circle with radius 1 inch, and have them draw arcs of length 2 and 2.5 (by "rolling" the ruler along the circle) and measuring the coordinates of the relevant terminal point. $t = 2$ should yield a terminal point at approximately $(-0.4, 0.9)$. $t = 2.5$ should yield a terminal point at approximately $(-0.8, 0.6)$.

▼ Examples

Some values of t, their reference numbers, and their terminal points.

t	$\bar{t}$	Terminal Point
$-\frac{33\pi}{2}$	$\frac{\pi}{2}$	$(0,-1)$
$\frac{55\pi}{4}$	$\frac{\pi}{4}$	$\left(\frac{\sqrt{2}}{2},-\frac{\sqrt{2}}{2}\right)$
$\frac{48\pi}{6}$	0	$(1,0)$
$\frac{16\pi}{3}$	$\frac{\pi}{3}$	$\left(-\frac{1}{2},-\frac{\sqrt{3}}{2}\right)$
10	$10-3\pi\approx 0.57522204$	$(-0.839,-0.544)$

▼ Group Work: The Reference Number

This activity allows students to discover a speedy method for finding reference numbers for some commonly encountered angles. After they are finished, ask them to go through and identify the quadrants of the relevant terminal points.

Answers:

1.

t	$\bar{t}$	Quadrant of Terminal Point
0	0	—
$\frac{\pi}{3}$	$\frac{\pi}{3}$	I
$\frac{2\pi}{3}$	$\frac{\pi}{3}$	II
$\frac{3\pi}{3}$	0	—
$\frac{4\pi}{3}$	$\frac{\pi}{3}$	III
$\frac{5\pi}{3}$	$\frac{\pi}{3}$	IV
$\frac{6\pi}{3}$	0	—
$\frac{7\pi}{3}$	$\frac{\pi}{3}$	I
$\frac{8\pi}{3}$	$\frac{\pi}{3}$	II
$\frac{9\pi}{3}$	0	—
$\frac{10\pi}{3}$	$\frac{\pi}{3}$	III
$\frac{11\pi}{3}$	$\frac{\pi}{3}$	IV
$\frac{35\pi}{3}$	$\frac{\pi}{3}$	IV
$\frac{1000\pi}{3}$	$\frac{\pi}{3}$	I

2.

t	$\bar{t}$	Quadrant of Terminal Point
0	0	—
$\frac{\pi}{6}$	$\frac{\pi}{6}$	I
$\frac{2\pi}{6}$	$\frac{\pi}{3}$	I
$\frac{3\pi}{6}$	$\frac{\pi}{2}$	—
$\frac{4\pi}{6}$	$\frac{\pi}{3}$	II
$\frac{5\pi}{6}$	$\frac{\pi}{6}$	II
$\frac{6\pi}{6}$	0	—
$\frac{7\pi}{6}$	$\frac{\pi}{6}$	III
$\frac{8\pi}{6}$	$\frac{\pi}{3}$	III
$\frac{9\pi}{6}$	$\frac{\pi}{2}$	—
$\frac{10\pi}{6}$	$\frac{\pi}{3}$	IV
$\frac{11\pi}{6}$	$\frac{\pi}{6}$	IV
$\frac{35\pi}{6}$	$\frac{\pi}{6}$	IV
$\frac{1000\pi}{6}$	$\frac{\pi}{3}$	II

▼ Homework Problems

Core Exercises: 3, 10, 21, 26, 30, 35, 40, 44, 54

Sample Assignment: 3, 6, 10, 13, 18, 21, 26, 30, 34, 35, 37, 40, 44, 49, 54, 55, 58

GROUP WORK, SECTION 2.1

The Reference Number

1. Find the reference numbers $\bar{t}$ of the following values of t.

t	$\bar{t}$
0	
$\frac{\pi}{3}$	
$\frac{2\pi}{3}$	
$\frac{3\pi}{3}$	
$\frac{4\pi}{3}$	
$\frac{5\pi}{3}$	
$\frac{6\pi}{3}$	
$\frac{7\pi}{3}$	
$\frac{8\pi}{3}$	
$\frac{9\pi}{3}$	
$\frac{10\pi}{3}$	
$\frac{11\pi}{3}$	
$\frac{35\pi}{3}$	
$\frac{1000\pi}{3}$	

2. Find the reference numbers $\bar{t}$ of the following values of t.

t	$\bar{t}$
0	
$\frac{\pi}{6}$	
$\frac{2\pi}{6}$	
$\frac{3\pi}{6}$	
$\frac{4\pi}{6}$	
$\frac{5\pi}{6}$	
$\frac{6\pi}{6}$	
$\frac{7\pi}{6}$	
$\frac{8\pi}{6}$	
$\frac{9\pi}{6}$	
$\frac{10\pi}{6}$	
$\frac{11\pi}{6}$	
$\frac{35\pi}{6}$	
$\frac{1000\pi}{6}$	

2.2 TRIGONOMETRIC FUNCTIONS OF REAL NUMBERS

▼ Suggested Time and Emphasis

$\frac{1}{2}$–1 class. Essential material.

▼ Points to Stress

1. Computation of the trigonometric functions, given the terminal point of a real number.
2. Properties of the trigonometric functions, including their domains, and parity based on the quadrant of the terminal point.
3. The Pythagorean and reciprocal identities of the trigonometric functions.

▼ Sample Questions

- **Text Question:** If $\sin t = \frac{3}{5}$, what is the value of $\csc t$?

  Answer: $\frac{5}{3}$
- **Drill Question:** If $\sin t = \frac{3}{5}$ and the terminal point of t is in the second quadrant, what is the value of $\cos t$?

  Answer: $-\sqrt{\frac{2}{5}}$

▼ In-Class Materials

- The etymology of the word sine is fairly interesting and not that well known. It starts with the Indian word *jya*, meaning "cord of a bowstring." The Arabs translated this word as *jiba*. The written language had no vowels, so it looked like this: *jb*. In 1145, the Spanish translator Robert of Cheste had to figure out what vowels to put in. Because of the shape of the curve, he thought the word was *jaib*, which meant the opening of a garment that shows a woman's cleavage. So he used the Latin word for the cavity formed by a curve: *sinus*. (This word exists in the English language today—right behind your nose are the sinus cavities.)
- There is an inconsistency in mathematical notation that can be made explicit at this time. We say that $\frac{1}{3} = 3^{-1}$. Similarly, we say that $\frac{1}{\pi} = \pi^{-1}$. But $\dfrac{1}{\sin x}$ is written as $\csc x$. Unfortunately, there is a symbol, $\sin^{-1} x$ that means something entirely different — the arcsine of x. Even worse, as noted in the section, $\sin^2 x = (\sin x)^2$ and $\sin^3 x = (\sin x)^3$. The exponent -1 is a notational anomaly.
- Problems such as Exercises 25–28 are key when students take calculus. It is vitally important that they go beyond memorizing the answers and truly understand where they come from.

▼ Examples

- $\sin \frac{\pi}{3} = \frac{\sqrt{3}}{2}$
- $\cos \frac{7\pi}{6} = -\frac{\sqrt{3}}{2}$
- $\tan \frac{11\pi}{4} = -1$
- $\sec \frac{17\pi}{3} = 2$
- $\csc \frac{17\pi}{2} = 1$
- $\cot \frac{121\pi}{6} = \sqrt{3}$

▼ Group Work: Fun with a Calculator

This activity foreshadows Section 2.3. Instruct the class to use their calculators to approximate values, but to draw the graphs by hand. Encourage them to use their group members effectively by dividing up the work in an efficient way. Give each group a different basic trigonometric function for Problem 2. If a group finishes early, give them a different function to try.

Answers:

1. Answers will vary.
2.

x	$\sin x$	$\cos x$	$\tan x$	$\csc x$	$\sec x$	$\cot x$
0	0	1	0	—	1	—
0.5	0.4794	0.8776	0.5463	2.0858	1.1395	1.8305
1	0.8415	0.5403	1.5574	1.1884	1.8508	0.6421
1.5	0.9975	0.0707	14.1014	1.0025	14.1368	0.0709
2	0.9093	−0.4161	−2.1850	1.0998	−2.4030	−0.4577
2.5	0.5985	−0.8011	−0.7470	1.6709	−1.2482	−1.3386
3	0.1411	−0.9900	−0.1425	7.0862	−1.0101	−7.0153
3.5	−0.3508	−0.9365	0.3746	−2.8508	−1.0679	2.6696
4	−0.7568	−0.6536	1.1578	−1.3213	−1.5299	0.8637
4.5	−0.9775	−0.2108	4.6373	−1.0230	−4.7439	0.2156
5	−0.9589	0.2837	−3.3805	−1.0428	3.5253	−0.2958
5.5	−0.7055	0.7087	−0.9956	−1.4174	1.4111	−1.0044
6	−0.2794	0.9602	−0.2910	−3.5789	1.0415	−3.4364
6.5	0.2151	0.9766	0.2203	4.6486	1.0240	4.5397
7	0.6570	0.7539	0.8714	1.5221	1.3264	1.1475
7.5	0.9380	0.3466	2.7060	1.0661	2.8849	0.3695
8	0.9894	−0.1455	−6.7997	1.0108	−6.8729	−0.1471
8.5	0.7985	−0.6020	−1.3264	1.2524	−1.6611	−0.7539
9	0.4121	−0.9111	−0.4523	2.4265	−1.0975	−2.2108
9.5	−0.0752	−0.9972	0.0754	−13.3065	−1.0028	13.2689
10	−0.5440	−0.8391	0.6484	−1.8382	−1.1918	1.5424
10.5	−0.8797	−0.4755	1.8499	−1.1368	−2.1029	0.5406
11	−1.0000	0.0044	−225.9508	−1.0000	225.9531	−0.0044
11.5	−0.8755	0.4833	−1.8114	−1.1423	2.0691	−0.5521
12	−0.5366	0.8439	−0.6359	−1.8637	1.1850	−1.5727
12.5	−0.0663	0.9978	−0.0665	−15.0780	1.0022	−15.0448
13	0.4202	0.9074	0.4630	2.3800	1.1020	2.1597
13.5	0.8038	0.5949	1.3511	1.2441	1.6809	0.7401
14	0.9906	0.1367	7.2446	1.0095	7.3133	0.1380
14.5	0.9349	−0.3549	−2.6341	1.0696	−2.8175	−0.3796
15	0.6503	−0.7597	−0.8560	1.5378	−1.3163	−1.1682
15.5	0.2065	−0.9785	−0.2110	4.8434	−1.0220	−4.7390
16	−0.2879	−0.9577	0.3006	−3.4734	−1.0442	3.3263

3.

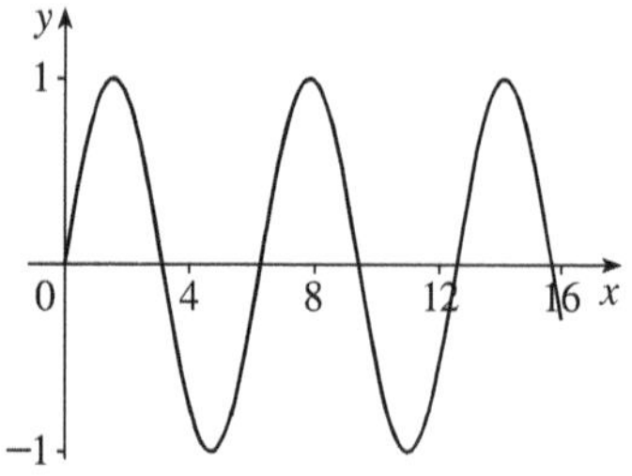

$\sin x$

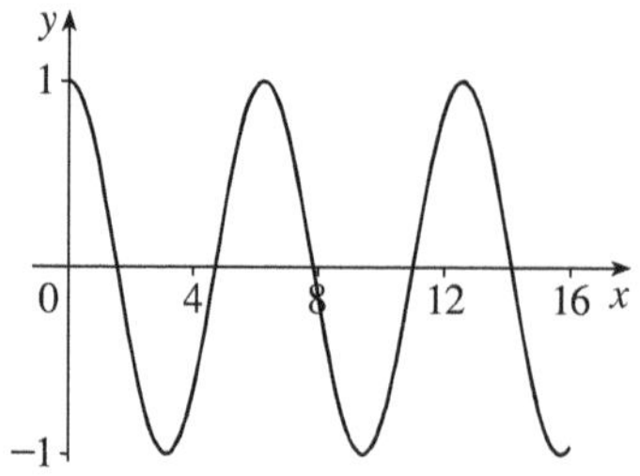

$\cos x$

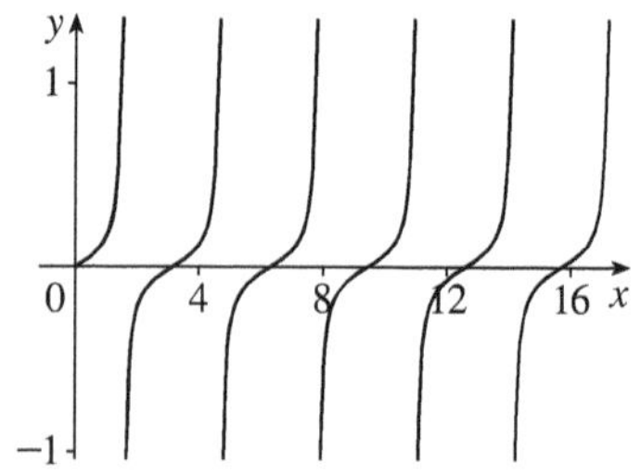

$\tan x$

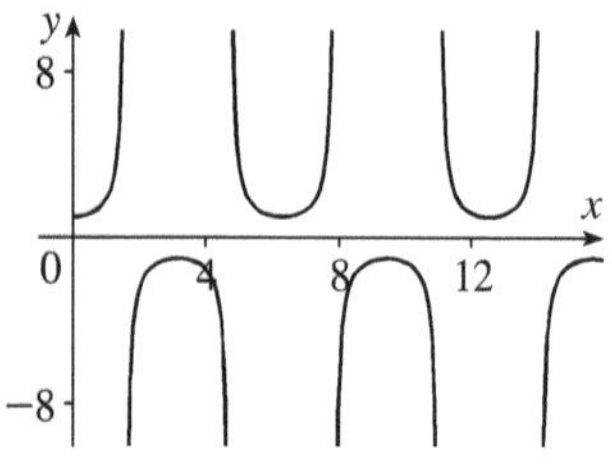

$\sec x$

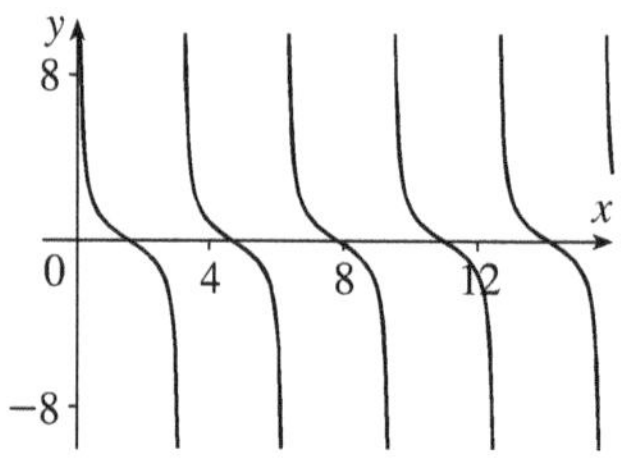

$\cot x$

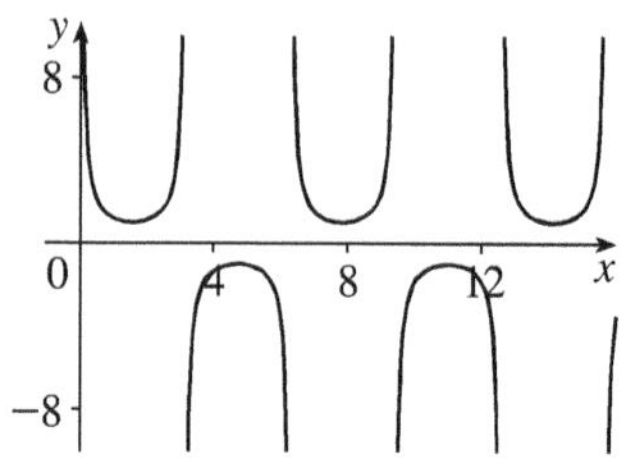

$\csc x$

4. For $\sin x$, $\cos x$, $\sec x$, and $\csc x$, 6.25 is a good estimate (actual value: 2π); for $\tan x$ and $\cot x$, 3 is a good estimate (actual value: π).

▼ Homework Problems

Core Exercises: 3, 5, 8, 15, 20, 30, 39, 51, 58, 66, 81

Sample Assignment: 3, 5, 8, 12, 15, 20, 23, 26, 30, 33, 39, 44, 48, 51, 58, 66, 73, 80, 81

GROUP WORK, SECTION 2.2

Fun with a Calculator

Consider the strange function graphed below:

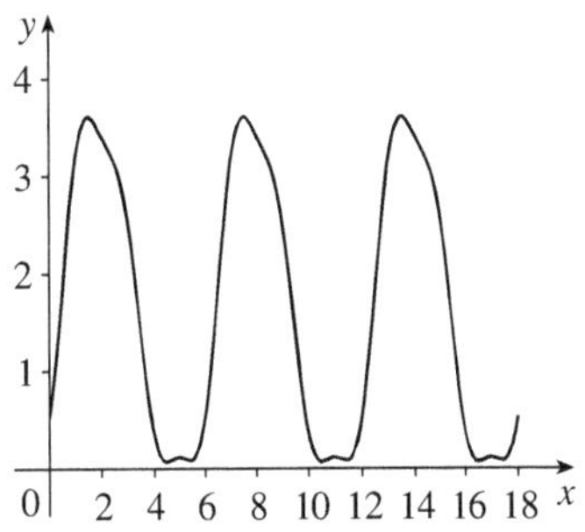

We say that the **period** of this function is 6.

1. In your own words, what is the period of a function?

2. Let $f(x) =$ _________. Fill in the following table of values:

x	$f(x)$
0	
0.5	
1	
1.5	
2	
2.5	
3	
3.5	
4	
4.5	
5	

x	$f(x)$
5.5	
6	
6.5	
7	
7.5	
8	
8.5	
9	
9.5	
10	
10.5	

x	$f(x)$
11	
11.5	
12	
12.5	
13	
13.5	
14	
14.5	
15	
15.5	
16	

3. Plot these points, and try to sketch the graph by connecting your points.

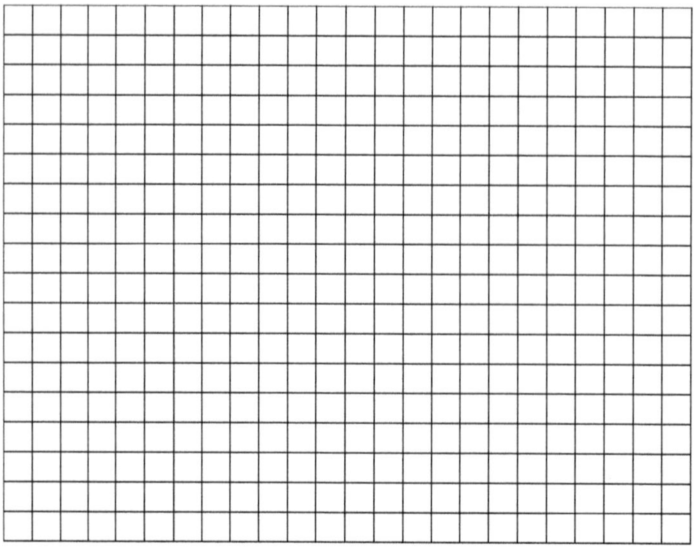

4. Estimate the period of your graph.

2.3 TRIGONOMETRIC GRAPHS

▼ Suggested Time and Emphasis

1 class. Essential material.

▼ Points to Stress

1. The graphs of sine and cosine, their periodicity, and their transformations.
2. Sine and cosine curves with variable amplitudes.

▼ Sample Questions

- **Text Question:** Sketch $y = \sin x$.

  Answer:

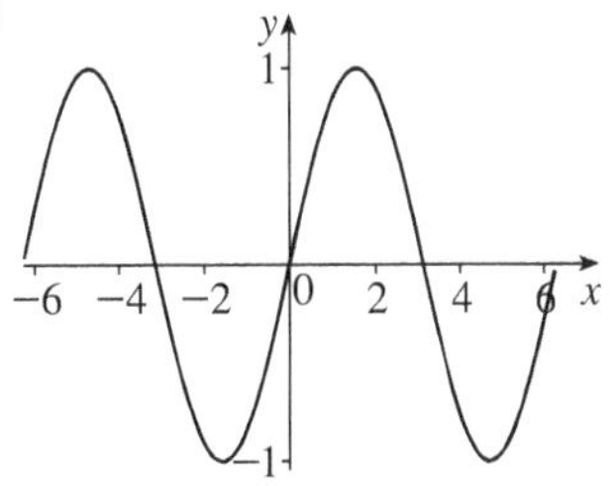

- **Drill Question:** Sketch $y = 2\sin \pi x$.

  Answer:

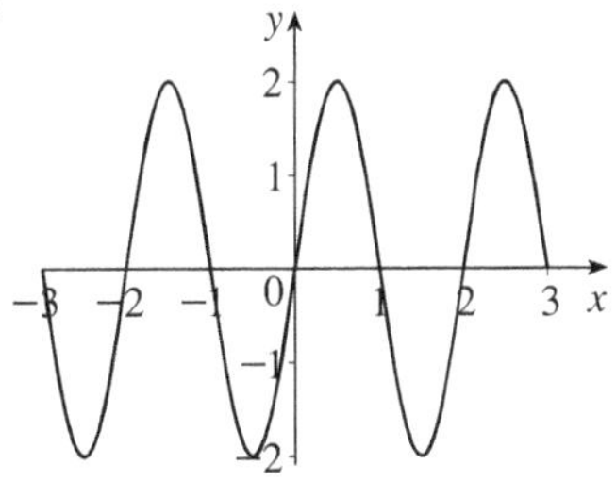

▼ In-Class Materials

- Let $f(x) = \sin x + \frac{1}{100}\cos x$. Have students graph this function with their calculators. In the default viewing window, the graph will be indistinguishable from that of $\sin x$. Show that when they zoom in they can see the "bumps" in the function.
- Demonstrate simple harmonic motion by bringing a spring (like a slinky, or another kind of spring, perhaps with a weight attached at the end) and letting it oscillate. Notice how the graph of the height of the end of the spring versus time is a sine (or cosine) curve. Similarly, if you chart the height with respect to time of a point on a spinning disk whose axis is parallel to the floor, you will get a sine or cosine curve.

- Ask the class if it is true that the sum of two periodic curves is always periodic. It turns out that although this seems a reasonable proposition, it is actually false. It is a true statement if you add the stipulation that the periods of the two curves are rational multiples of each other.

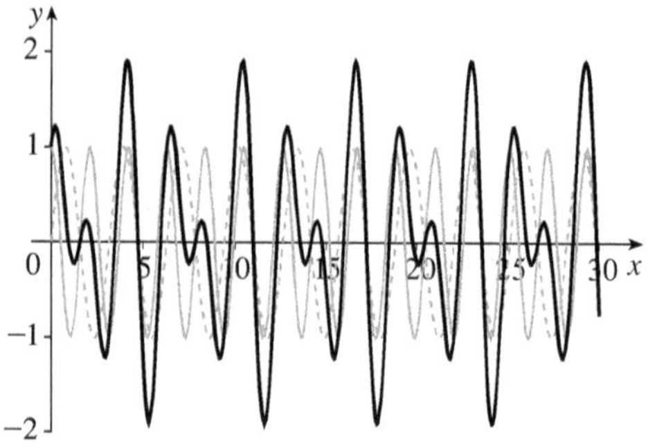

$\sin 2x + \cos 3x$ is periodic

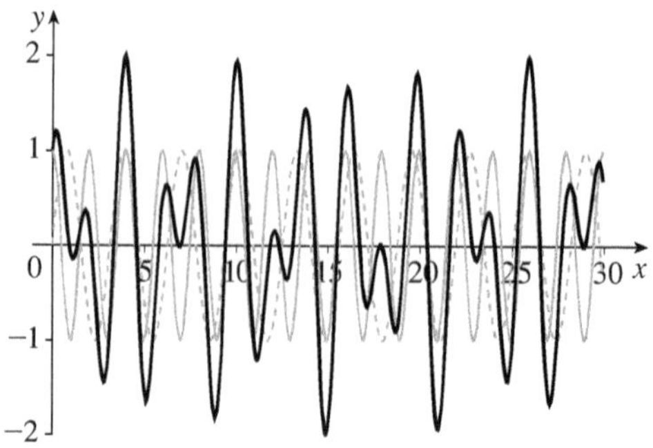

$\sin 2x + \cos \pi x$ is not periodic

- Have students obtain a table of time of sunrise versus date for your city. Perhaps have each student write down the information for a given month. (Websites such as `http://usno.navy.mil` will do these calculations for you.) Have the students pool and plot the data, and see how the resulting graph resembles a sine or cosine curve.

▼ Examples

Graphs of variants of sine and cosine curves:

- $y = 3\sin 2x + 1$ has amplitude 3, period π, and vertical shift 1.

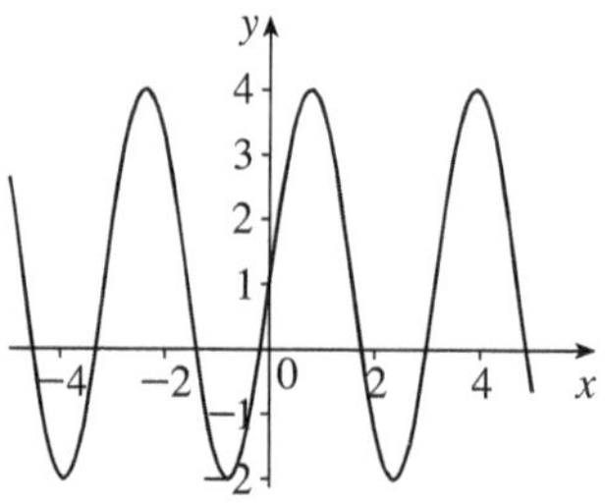

- $y = \frac{1}{2}\cos\left(\frac{\pi}{3}(x - 1)\right)$ has amplitude $\frac{1}{2}$, period 6, and horizontal shift 1.

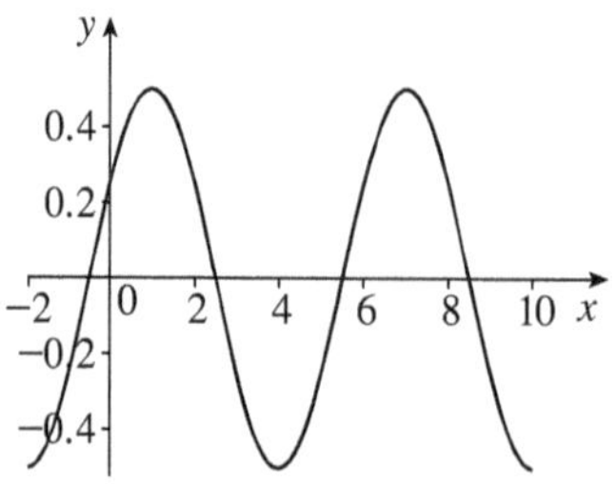

▼ Group Work 1: Pushing the Envelope

This is an extension of Examples 8 and 9. Start the activity by discussing these examples with students. This activity serves as a review of earlier sections, requiring students to recall what the graphs of exponential, logarithmic and polynomial functions look like. It should be done without a calculator.

Answers:

1. I = (b), II = (a), III = (d), IV = (c), V = (h), VI = (e), VII = (f), VIII = (g)
2. I = (c), II = (h), III = (a), IV = (f), V = (e), VI = (d), VII = (b), VIII = (g)

▼ Group Work 2: Which is the Original?

This activity allows students to experiment with such concepts as amplitude, period, and shifts of the graph of the sine function. If a group finishes early, ask them to sketch the graphs of $f(x)+2$, $f(x)-2$, $f(x+2)$, and $f(x-2)$.

Answers:

1. $2f(x)$ **2.** $\frac{1}{2}f(2(x-3))$ **3.** $f(x)$ **4.** $\frac{1}{2}f(x)$ **5.** $f(x-1)$ **6.** $f\left(\frac{1}{2}x\right)$ **7.** $2f\left(\frac{1}{2}x\right)$ **8.** $f(2x)$

▼ Group Work 3: Fun with Fourier

This activity should be given before Fourier series are discussed in class. This activity will get students looking at combinations of sine curves, while at the same time foreshadowing the concepts of infinite series and Fourier series.

Answers:

1. No

2. $S(x)=\begin{cases}1 & \text{if } -2\pi\le x<-\pi \text{ or } 0\le x<\pi\\ -1 & \text{if } -\pi\le x<0 \text{ or } \pi\le x\le 2\pi\end{cases}$

3. $\sin x$, because at least it is an odd function, like the square wave.

4. $\frac{4}{\pi}\sin x$

5. $\frac{4}{\pi}\left(\sin x+\frac{1}{3}\sin 3x+\frac{1}{5}\sin 5x\right)$

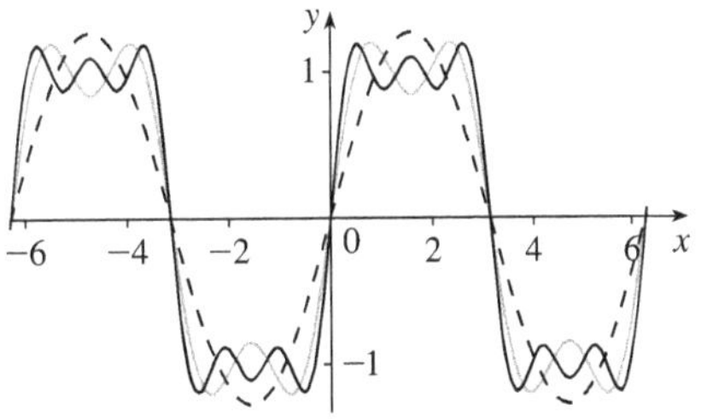

6. $\frac{4}{\pi}\left(\sin x+\frac{\sin 3x}{3}+\frac{\sin 5x}{5}+\frac{\sin 7x}{7}+\frac{\sin 9x}{9}+\frac{\sin 11x}{11}+\frac{\sin 13x}{13}+\frac{\sin 15x}{15}+\frac{\sin 17x}{17}+\frac{\sin 19x}{19}\right)$

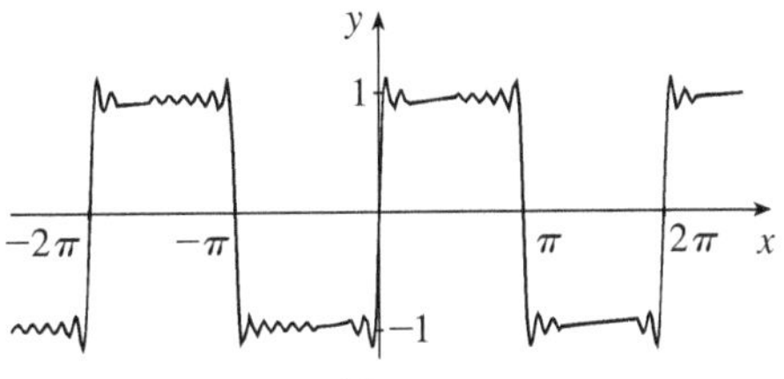

7.

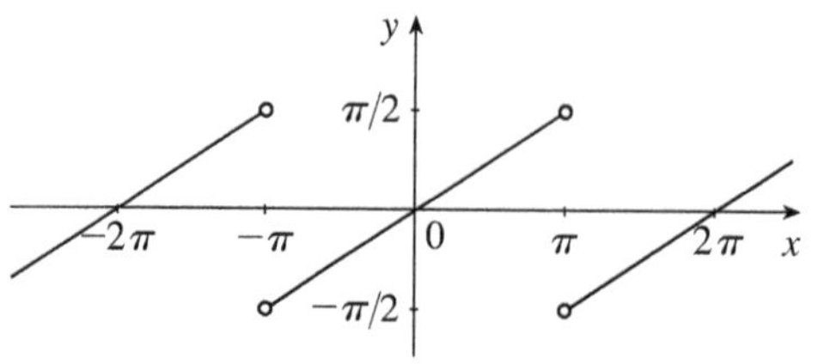

▼ Homework Problems

Core Exercises: 7, 20, 29, 46, 61, 68, 75, 78

Sample Assignment: 7, 14, 20, 27, 29, 34, 46, 50, 61, 65, 68, 74, 75, 78, 82

Pushing the Envelope

Match the following labels to their graphs.

1. I $e^x \sin 10x$ II $e^x \cos 10x$ III $e^{-x} \sin 10x$ IV $e^{-x} \cos 10x$
V $e^{-2x} \sin 10x$ VI $e^{-2x} \cos 10x$ VII $e^{2x} \sin 10x$ VIII $e^{2x} \cos 10x$

(a)

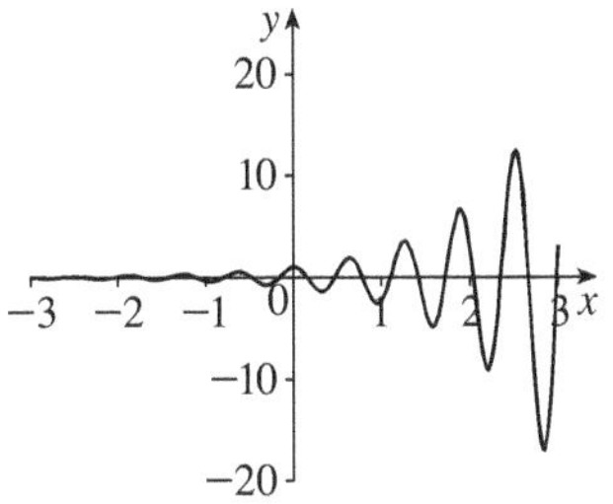

(b)

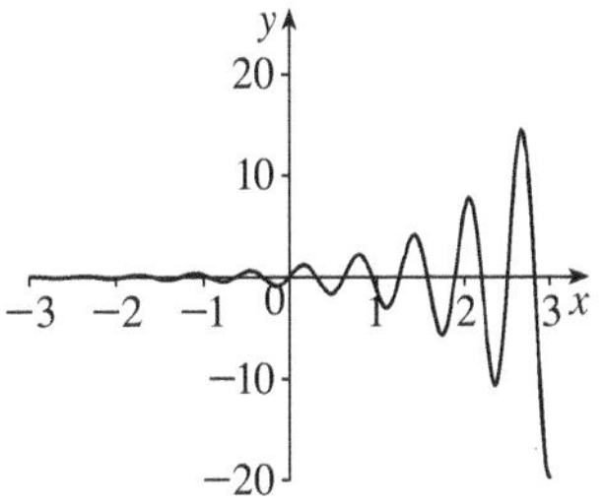

(c)

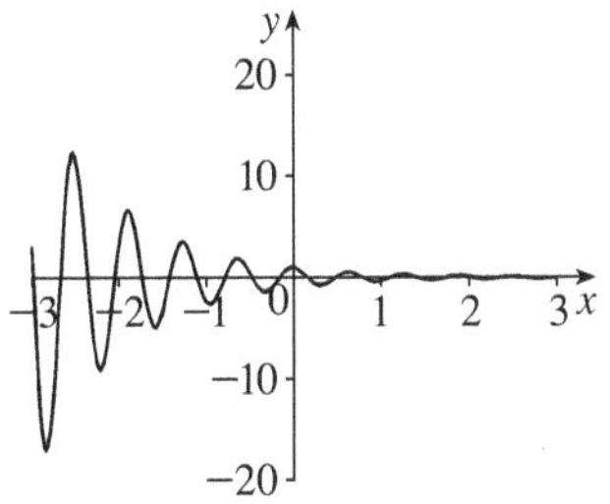

(d)

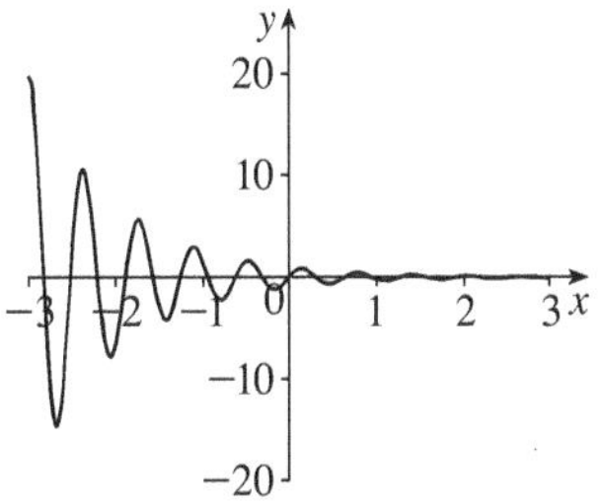

(e)

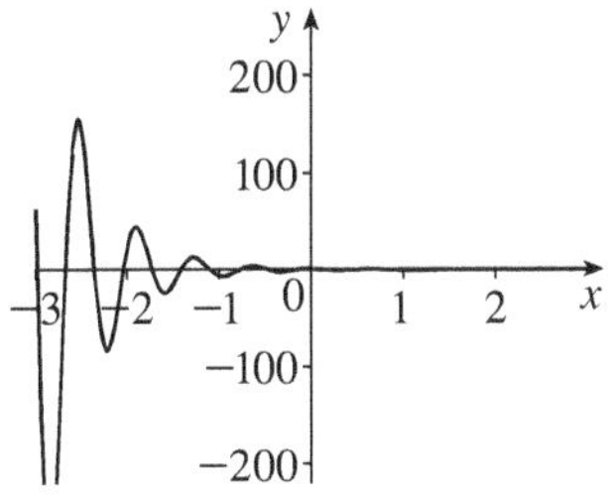

(f)

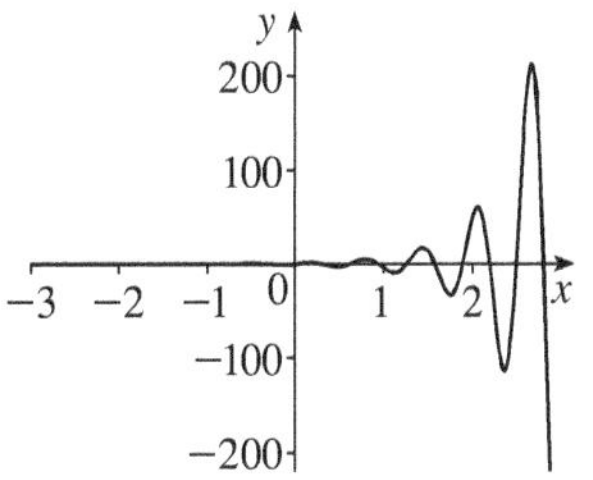

(g)

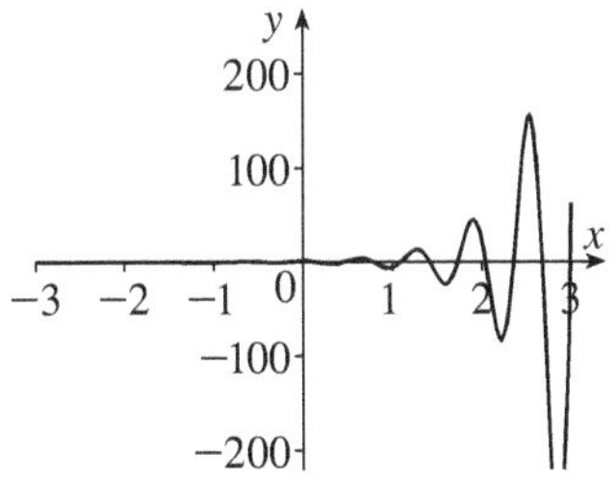

(h)

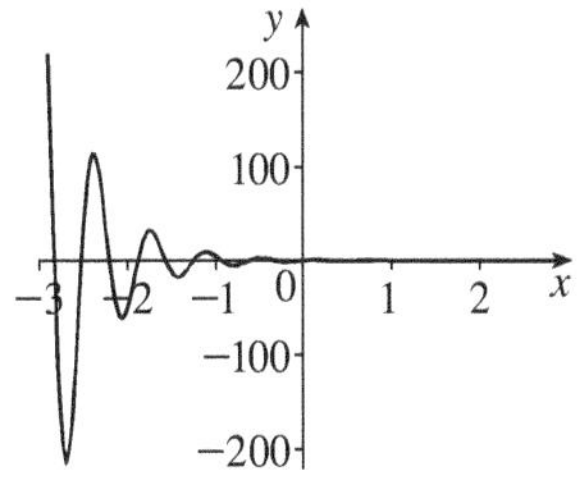

2. I $\ln x \sin 20x$ II $x^2 \sin 20x$ III $-x^2 \sin 20x$ IV $\cos x \sin 20x$
V $(3 - x^2) \sin 20x$ VI $\sin \frac{1}{2}x \sin 20x$ VII $\frac{1}{x} \sin 20x$ VIII $\sqrt{x} \sin 20x$

(a)

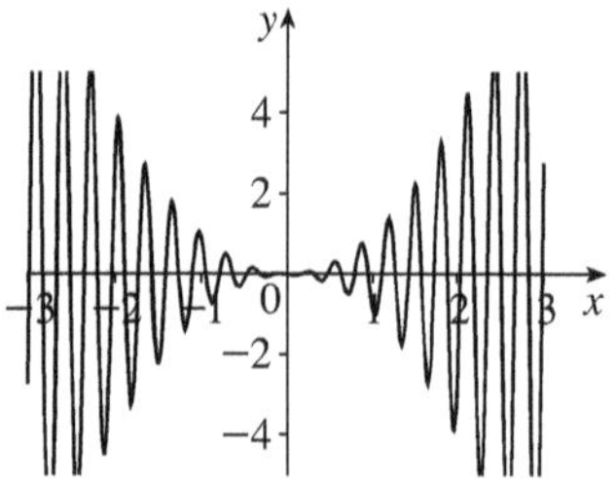

(b)

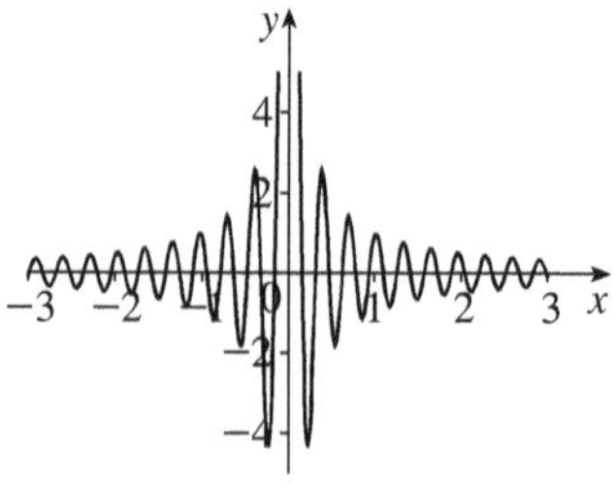

(c)

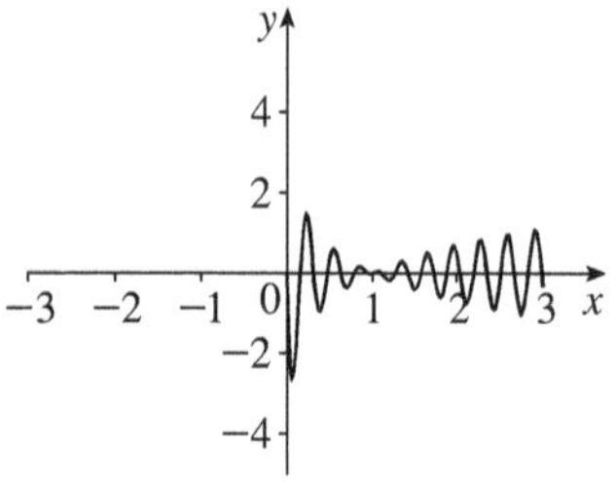

(d)

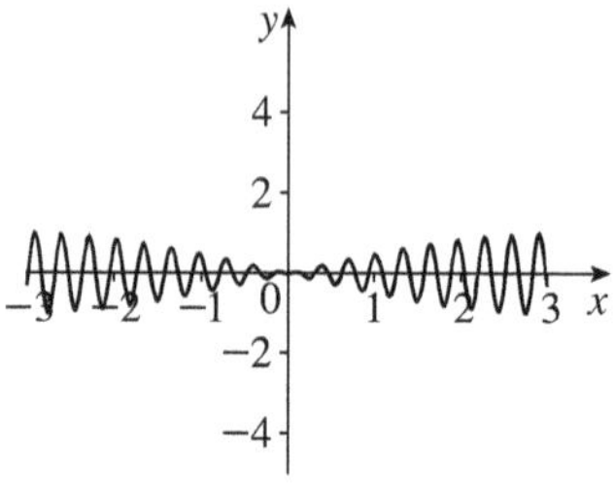

(e)

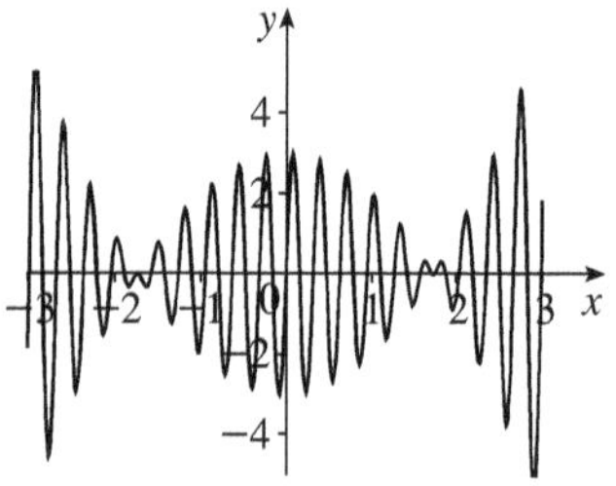

(f)

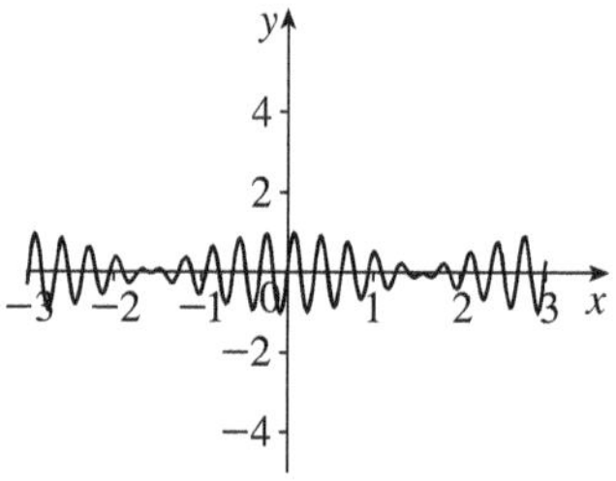

(g)

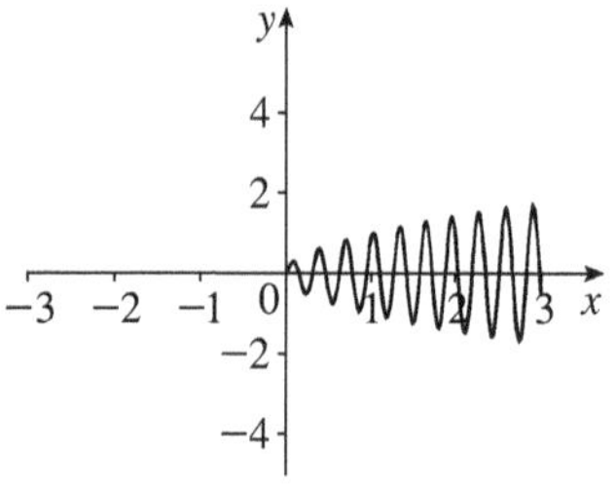

(h)

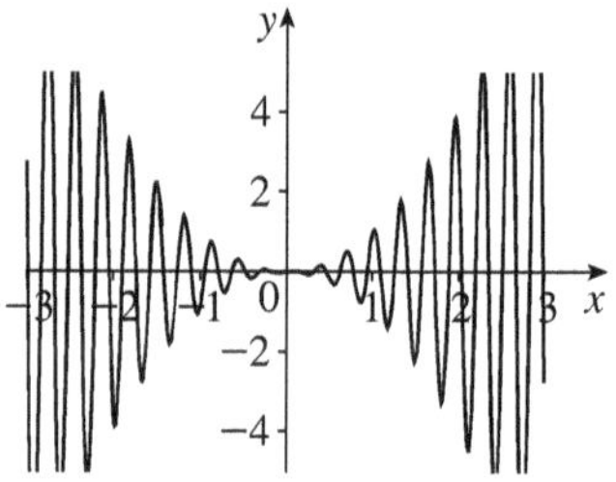

GROUP WORK 2, SECTION 2.3

Which is the Original?

You see before you graphs of $f(x)$, $2f(x)$, $f(x-1)$, $\frac{1}{2}f(x)$, $f(2x)$, $2f\left(\frac{1}{2}x\right)$, $f\left(\frac{1}{2}x\right)$, and $\frac{1}{2}f(2(x-3))$. Give each graph the appropriate label.

1.

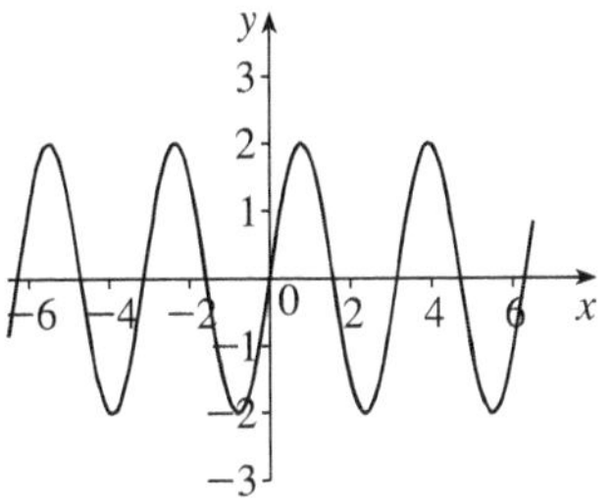

2.

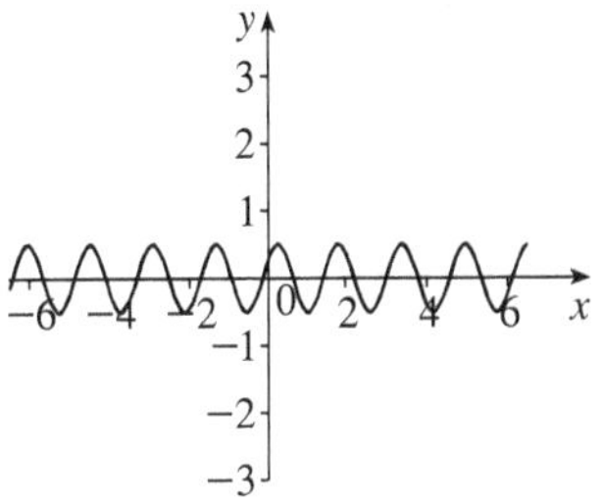

3.

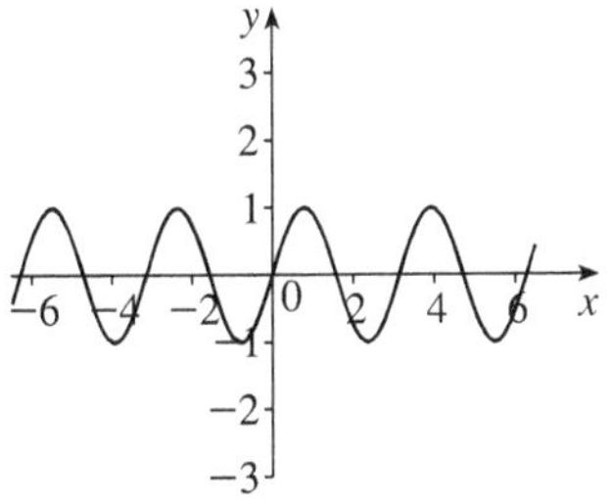

4.

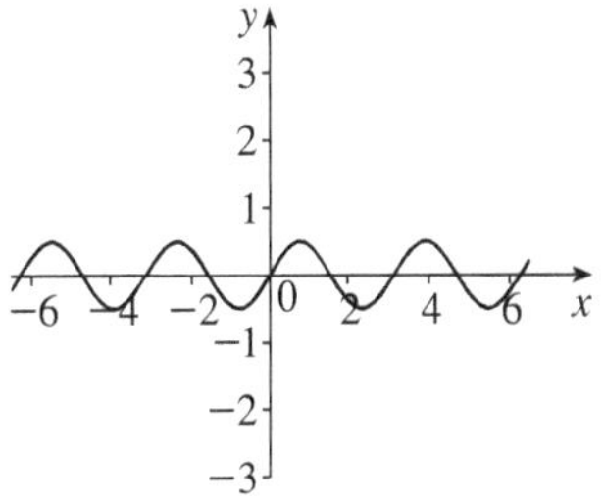

5.

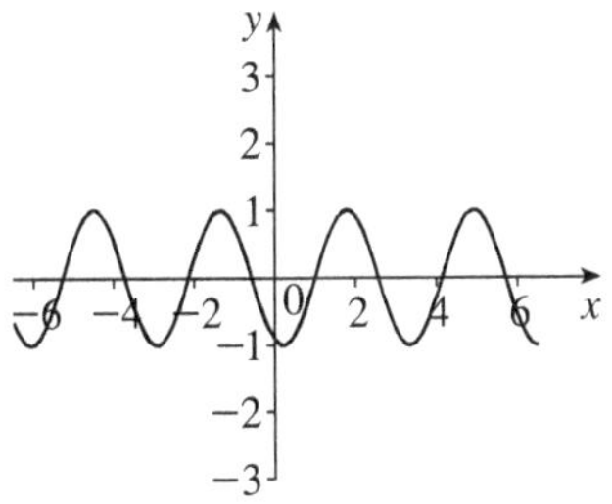

6.

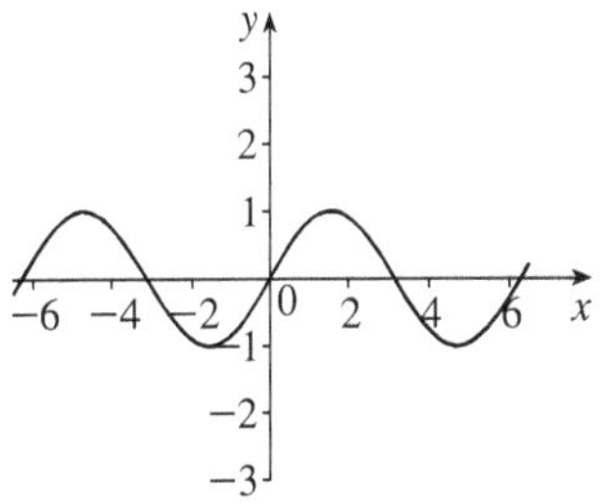

7.

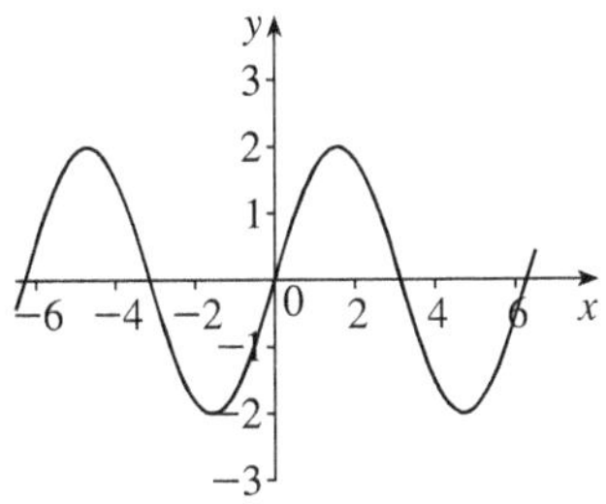

8.

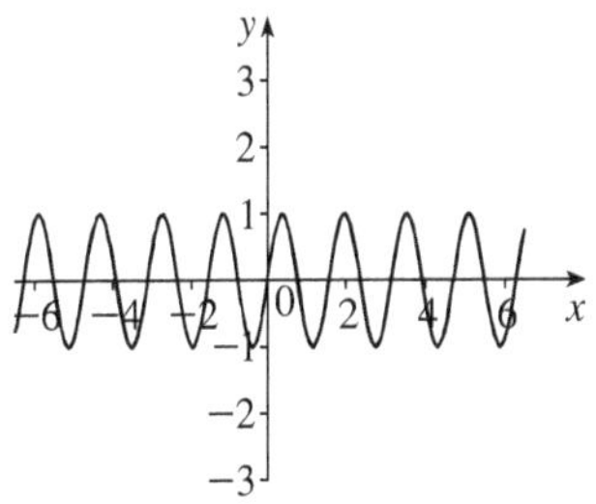

GROUP WORK 3, SECTION 2.3

Fun with Fourier

The following function $S(x)$ is called a **square wave**.

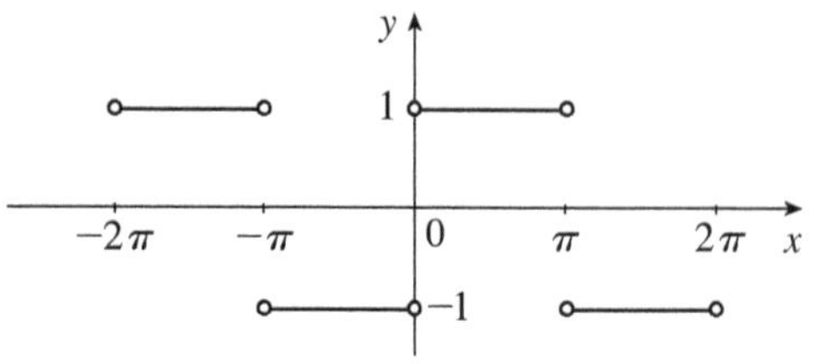

1. Can you find a function on your calculator which has the given graph?

2. Write a formula that has this graph for $-2\pi \le x \le 2\pi$.

3. Which of the functions $\sin x$ and $\cos x$ gives a better approximation of $S(x)$?

4. Which of $\frac{4}{\pi}\sin x$ and $\frac{4}{\pi}\cos x$ gives a better approximation of $S(x)$?

5. Graph $\frac{4}{\pi}\sin x$, $\frac{4}{\pi}\left(\sin x + \frac{1}{3}\sin 3x\right)$, and $\frac{4}{\pi}\left(\sin x + \frac{1}{3}\sin 3x + \frac{1}{5}\sin 5x\right)$. Which of these functions gives the best approximation of $S(x)$?

6. A **Fourier approximation** of a function is an approximation of the form

$$F(x) = a_0 + a_1 \cos x + b_1 \sin x + a_2 \cos 2x + b_2 \sin 2x + \cdots + a_n \cos nx + b_n \sin nx$$

You have just discovered the Fourier approximation to $S(x)$with three terms. Find the Fourier approximation to $S(x)$with ten terms, and sketch its graph.

7. The following expressions are Fourier approximations to a different function, $T(x)$:

$$\begin{aligned} T(x) &\approx \sin x \\ T(x) &\approx \sin x - \tfrac{1}{2}\sin 2x \\ T(x) &\approx \sin x - \tfrac{1}{2}\sin 2x + \tfrac{1}{3}\sin 3x \\ T(x) &\approx \sin x - \tfrac{1}{2}\sin 2x + \tfrac{1}{3}\sin 3x - \tfrac{1}{4}\sin 4x \\ T(x) &\approx \sin x - \tfrac{1}{2}\sin 2x + \tfrac{1}{3}\sin 3x - \tfrac{1}{4}\sin 4x + \tfrac{1}{5}\sin 5x \end{aligned}$$

Sketch $T(x)$.

2.4 MORE TRIGONOMETRIC GRAPHS

▼ Suggested Time and Emphasis

$\frac{1}{2}$ class. Essential material.

▼ Point to Stress

The graphs of tangent, cotangent, secant, and cosecant, their periodicity, and their transformations.

▼ Sample Questions

- **Text Question:** Sketch $y = \tan x$.

  Answer:

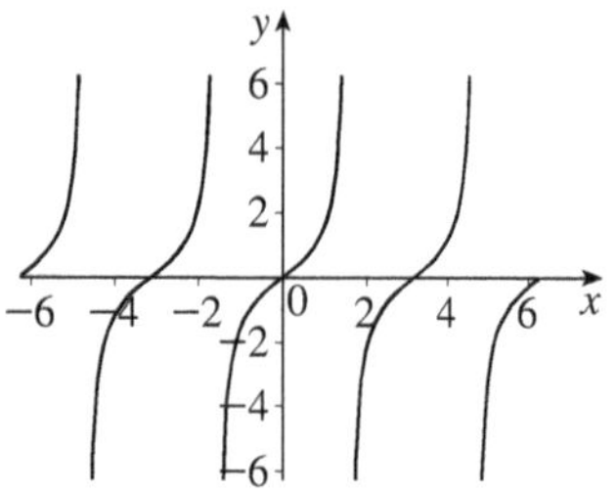

- **Drill Question:** Sketch $y = \tan \pi x$.

  Answer:

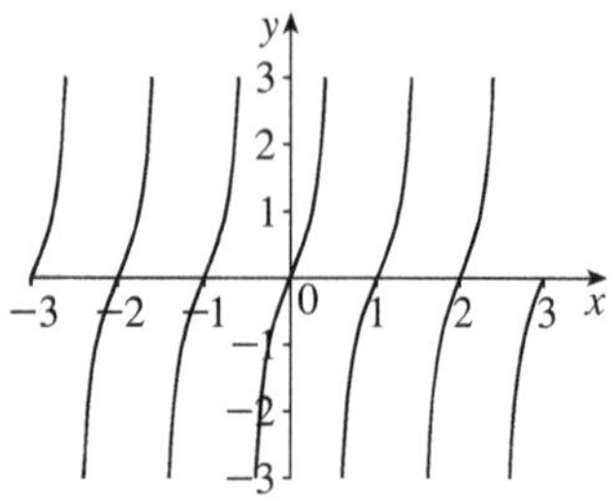

▼ In-Class Materials

- Notice that the functions $\sec x$ and $x^2 + 1$ are very different functions. ($\sec x$ has vertical asymptotes, for example, and $x^2 + 1$ does not.) When we casually sketch their graphs near $x = 0$, however, the sketches look similar:

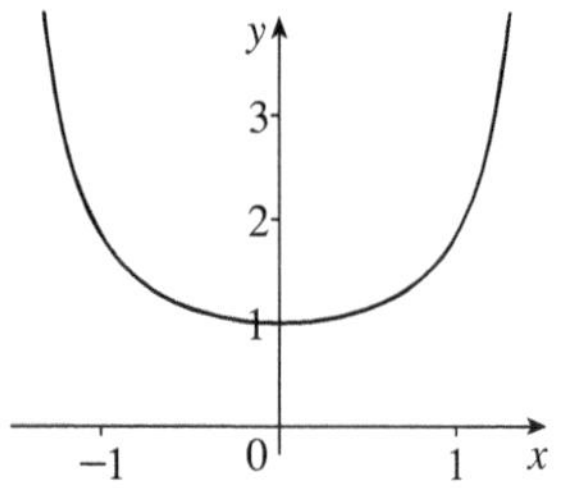

$y = \sec x$

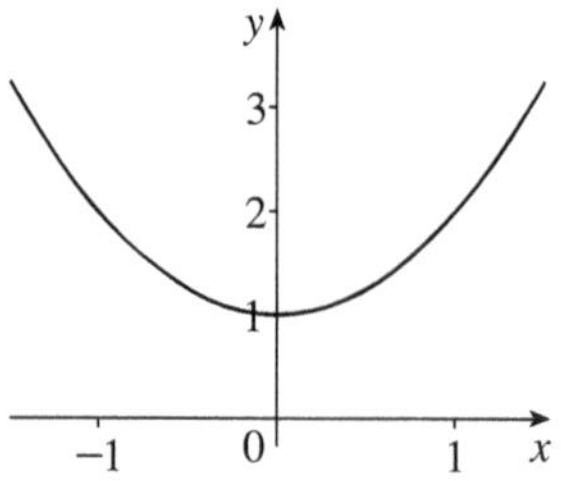

$y = x^2 + 1$

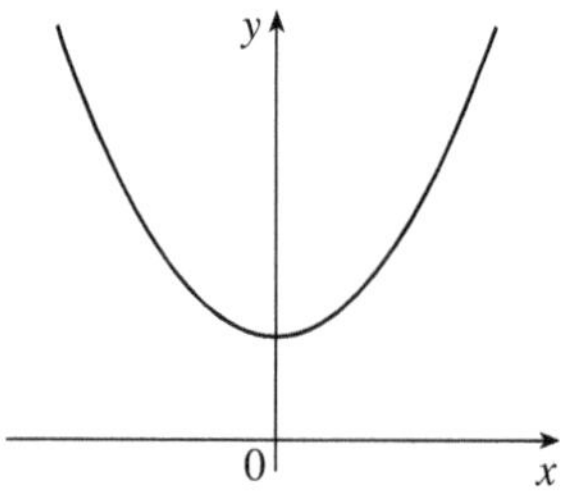

A typical sketch

The problem is that a lot of curves have a "U" shape near zero. Have students try to come up with other examples. If they have calculators, they can experiment with functions they haven't learned about yet.

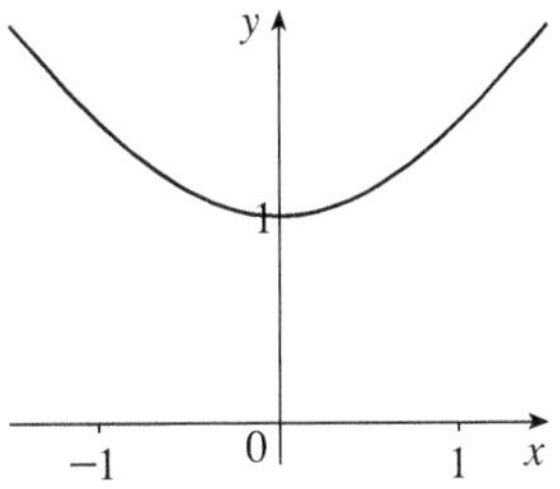

$y = -\cos x + 2$

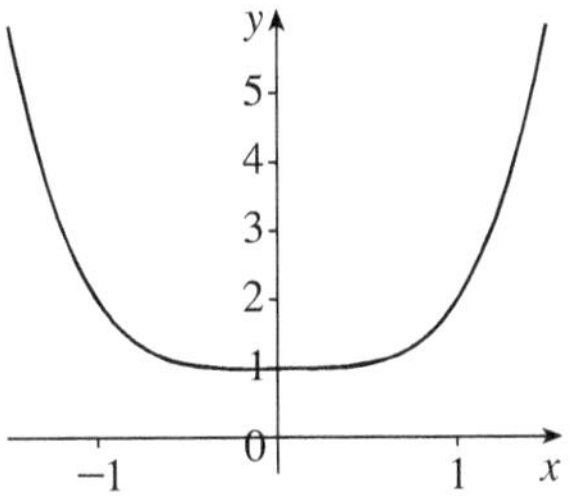

$y = x^4 + 1$

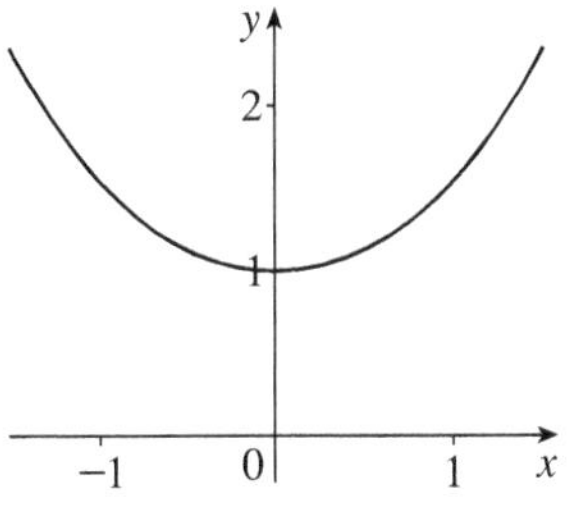

$y = \cosh x$

Of course, if we look at these functions' behavior on a larger interval, their similarity decreases, but some of them would still be hard to tell apart just by looking at the shapes of the curves.

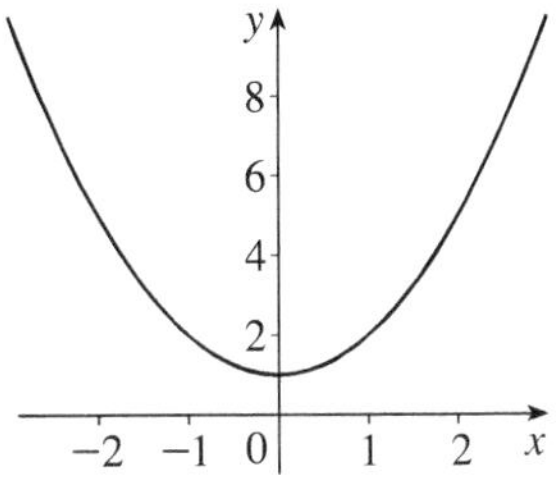

$y = x^2 + 1$

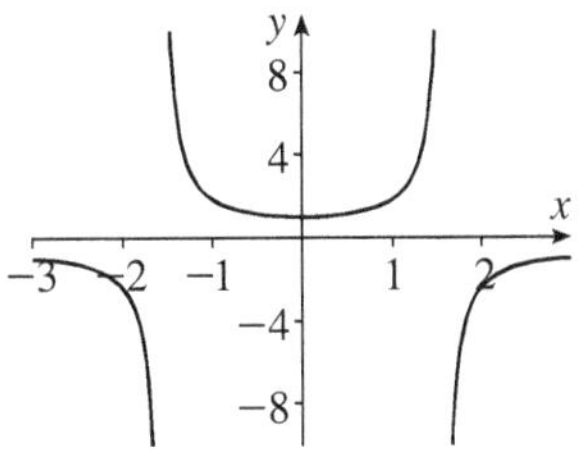

$y = \sec x$

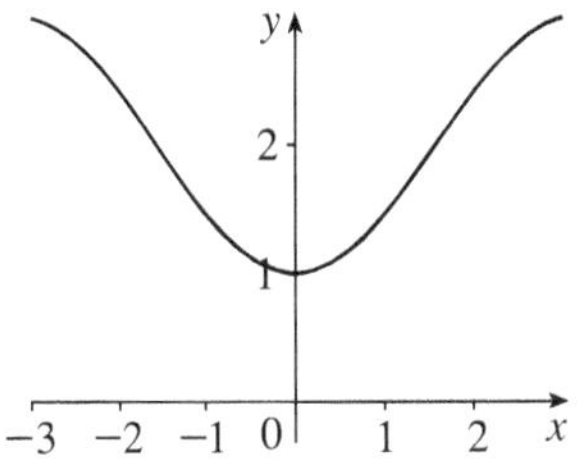

$y = -\cos x + 2$

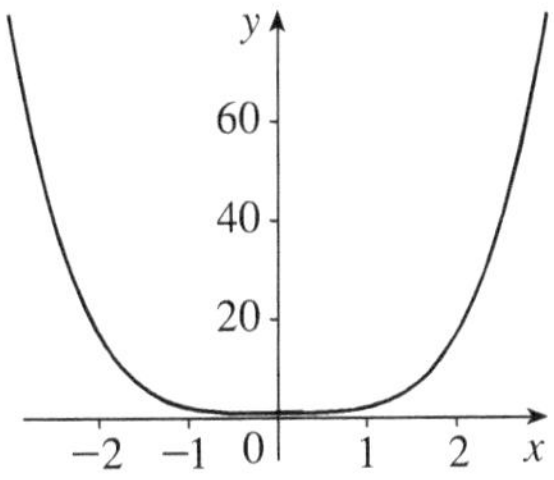

$y = x^4 + 1$

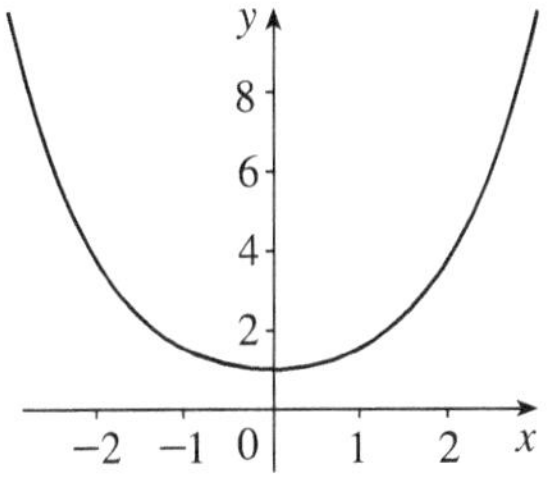

$y = \cosh x$

- Similarly, discuss $\tan x$, x^3, $\sin x$, and $\sinh x$. Notice that, unlike in the previous example, these curves don't all look alike at the origin — some are flatter than others.

▼ Examples

Graphs of variants of trigonometric curves:

- $y = \tan 2x + 1$ and $y = \cot 2x + 1$ have period $\frac{\pi}{2}$ and vertical shift 1.

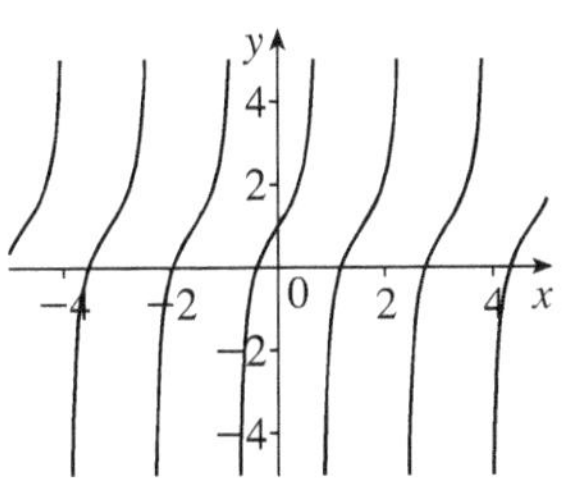

$y = \tan 2x + 1$

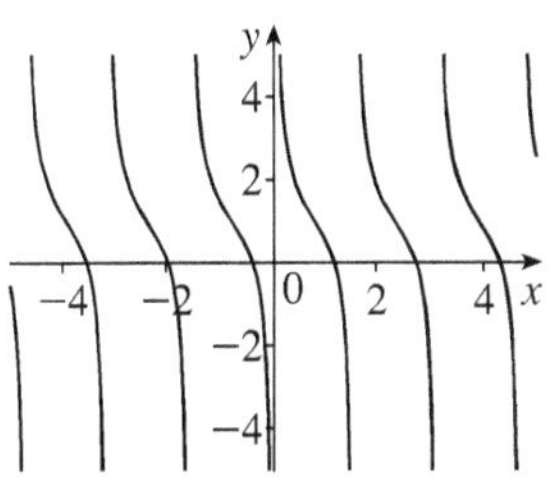

$y = \cot 2x + 1$

- $y = \sec\left(\frac{\pi}{3}(x-1)\right)$ and $y = \csc\left(\frac{\pi}{3}(x-1)\right)$ have period 3 and horizontal shift 1.

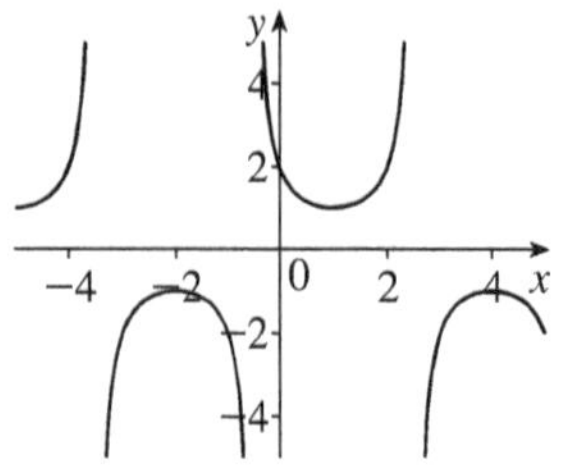

$y = \sec\left(\frac{\pi}{3}(x-1)\right)$

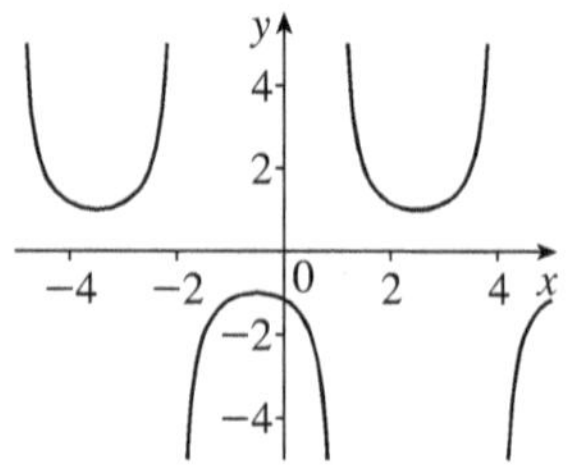

$y = \csc\left(\frac{\pi}{3}(x-1)\right)$

▼ Group Work: More Trigonometric Graphs

This exercise assumes that students have access to graphing calculators or some other form of graphing technology.

Answers: Answers will vary. Students should obtain pictures like the following:

1.

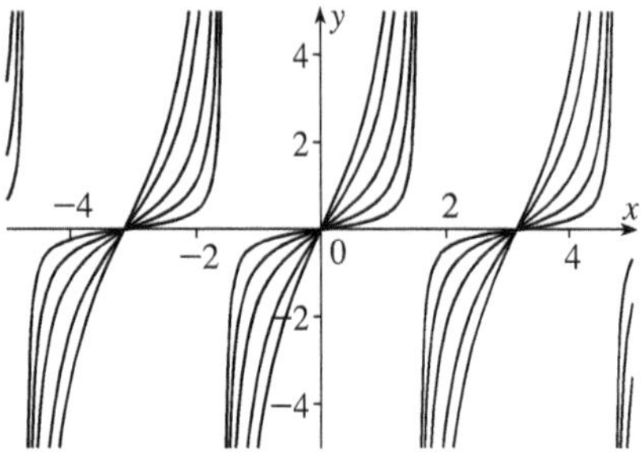

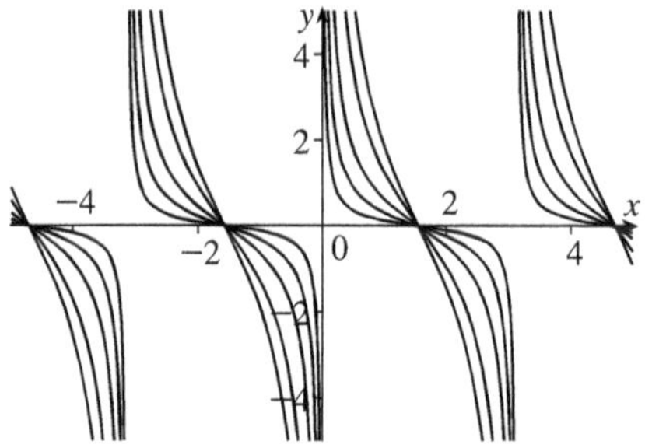

2.

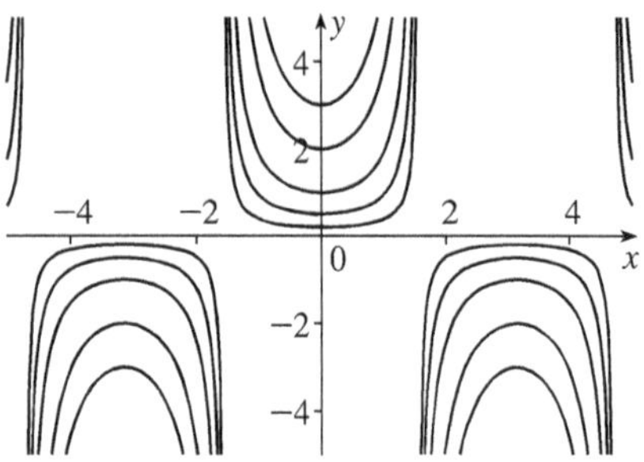

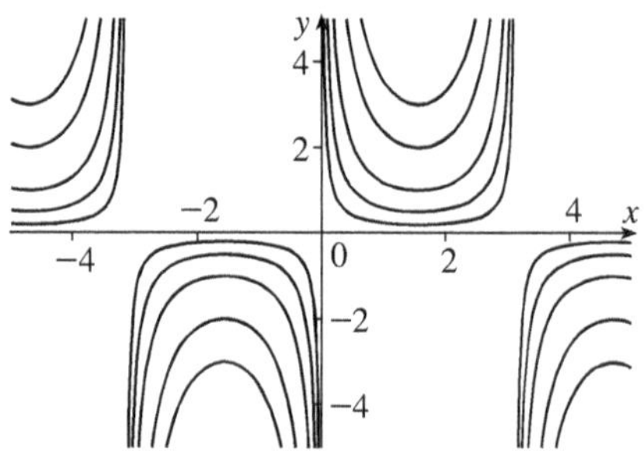

▼ Homework Problems

Core Exercises: 1, 4, 8, 13, 26, 38, 51, 55

Sample Assignment: 1, 4, 8, 13, 19, 26, 28, 38, 43, 50, 51, 55, 58

GROUP WORK, SECTION 2.4

More Trigonometric Graphs

It isn't that hard to understand the amplitude of sine and cosine graphs. For example, make sure everyone in your group remembers the difference between $y = \sin x$ and $y = 2\sin x$. Take a moment to remind each other what k does in $y = k\sin x$. Go ahead and do it. I will wait.

It is a little more difficult to see the role of k in these functions:

$$\begin{aligned} y &= k\tan x \\ y &= k\cot x \\ y &= k\sec x \\ y &= k\csc x \end{aligned}$$

1. We want to understand the graph of $y = k\tan x$. You won't be able to enter that expression in your calculators, but you can enter $y = \tan x$, $y = 2\tan x$, $y = 3\tan x$, and so forth. Try graphing $y = k\tan x$ for various values of k, all on the same set of axes. Don't forget to play with negative values of k, values of k that are less than 1, and values of k that are much less than 1. In your own words, describe what happens to $y = k\tan x$ for various values of k.

2. Repeat Problem 1 for the other three functions mentioned above.

2.5 INVERSE TRIGONOMETRIC FUNCTIONS AND THEIR GRAPHS

▼ Suggested Time and Emphasis

1 classes. Essential material.

▼ Points to Stress

1. Defining the inverse trigonometric functions.
2. Evaluating the inverse trigonometric functions.
3. Graphing the inverse trigonometric functions.

▼ Sample Questions

- **Text Question:** We can say $\sin^{-1} y = x$ means $y = \sin x$. Given this nice definition, why does the text need to fuss with domains and ranges?

  Answer: For a given y between -1 and 1, there are infinitely many values of x such that $y = \sin x$.
- **Drill Question:** Compute $\sin^{-1}\left(-\frac{1}{2}\right)$ and $\cos^{-1}\left(-\frac{\sqrt{3}}{2}\right)$.

  Answer: $-\frac{\pi}{6}, \frac{5\pi}{6}$

▼ In-Class Materials

- It may be helpful to start by reviewing general inverse functions (Section 1.8). If $f(x) = x^3$, what is $f^{-1}(x)$? Note that $f(2) = 8$ and $f^{-1}(8) = 2$. Now move to $f(x) = \sqrt[5]{3x+2} \Rightarrow f^{-1}(x) = \dfrac{x^5 - 2}{3}$. Note that $f(5) \approx 1.762$ and $f^{-1}(1.762) \approx 5$. Now explore $f(x) = x^2$, and point out that even though there is no inverse function, we could make things work out if we were allowed to assume that x was positive. This leads to a natural segue into $f(x) = \sin x$.
- Many students may need a review of notation: $\sin^2 x = (\sin x)^2$, $\sin x^2 = \sin(x^2)$, $(\sin x)^{-1} = \dfrac{1}{\sin x} = \csc x$, $\sin x^{-1} = \sin\left(\dfrac{1}{x}\right)$, but $\sin^{-1} x$ represents the inverse sine of x, the arcsine of x, which is not the same as any of the previous functions.
- Sketch the graph of the non-function $x = \sin y$. Demonstrate how a range restriction makes this a function, and why a range restriction on this function corresponds to a domain restriction on sine.

▼ Examples

- $\sin^{-1}\left(\frac{\sqrt{2}}{2}\right) = \frac{\pi}{4}$
- $\cos^{-1}\left(-\frac{\sqrt{3}}{2}\right) = \frac{5\pi}{6}$
- $\tan^{-1}\sqrt{3} = \frac{\pi}{3}$
- $\tan^{-1}\left(-\sqrt{3}\right) = -\frac{\pi}{3}$
- $\tan\left(\sin^{-1}\frac{1}{5}\right) = \frac{\sqrt{6}}{12}$
- $\cos\left(\tan^{-1}\frac{3}{8}\right) = \frac{8\sqrt{73}}{73}$
- $\sin^{-1}\left(\sin\frac{87\pi}{4}\right) = -\frac{\pi}{4}$

▼ Group Work: The Triangle Wave

Encourage students to divide up the work in Problem 1; the calculations are fairly routine.

The triangle wave is important in waveform analysis. It is the waveform of a rich, brassy sound.

Answers:

1. $\sin^{-1}(\sin 1) = 1$
   $\sin^{-1}(\sin 2) = -2 + \pi \approx 1.1415927$
   $\sin^{-1}(\sin 3) = -3 + \pi \approx 0.14159265$
   $\sin^{-1}(\sin 4) = -4 + \pi \approx -0.85840735$
   $\sin^{-1}(\sin 5) = 5 - 2\pi \approx -1.2831853$
   $\sin^{-1}(\sin 6) = 6 - 2\pi \approx -0.28318531$
   $\sin^{-1}(\sin 7) = 7 - 2\pi \approx 0.71681469$
   $\sin^{-1}(\sin 8) = -8 + 3\pi \approx 1.424778$
   $\sin^{-1}(\sin 9) = -9 + 3\pi \approx 0.42477796$
   $\sin^{-1}(\sin 10) = -10 + 3\pi \approx -0.57522204$

2. 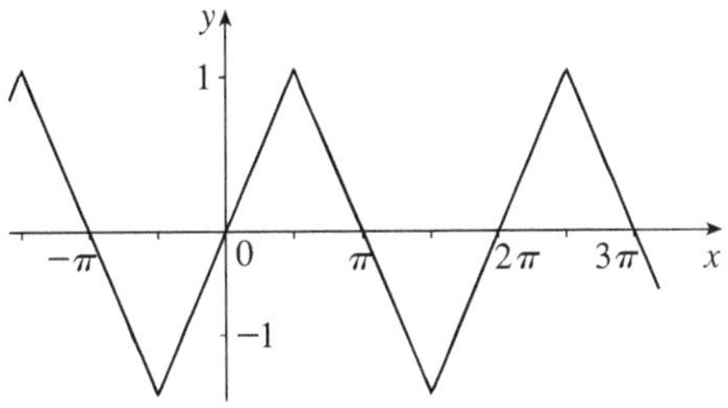

3. 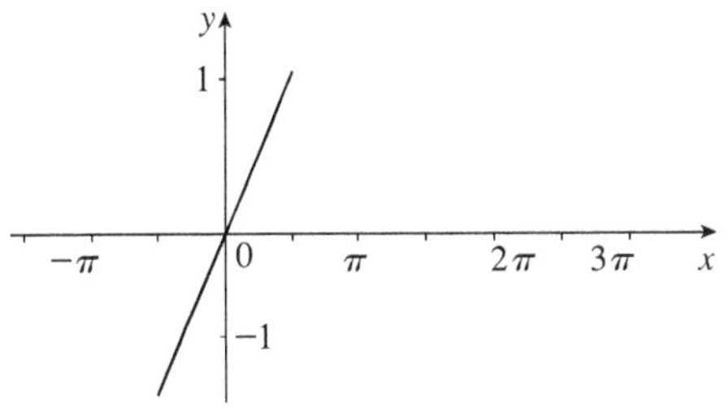

They are not the same because $\sin^{-1} x$ is defined only for $x \in [-1, 1]$.

▼ Homework Problems

Core Exercises: 2, 9, 19, 32, 44, 56, 57

Sample Assignment: 2, 4, 6, 9, 13, 19, 21, 28, 32, 38, 42, 44, 49, 56, 57, 59

GROUP WORK, SECTION 2.5

The Triangle Wave

1. Compute the following:

$\sin^{-1}(\sin 1)$	$\sin^{-1}(\sin 6)$
$\sin^{-1}(\sin 2)$	$\sin^{-1}(\sin 7)$
$\sin^{-1}(\sin 3)$	$\sin^{-1}(\sin 8)$
$\sin^{-1}(\sin 4)$	$\sin^{-1}(\sin 9)$
$\sin^{-1}(\sin 5)$	$\sin^{-1}(\sin 10)$

2. Let $f(x) = \sin^{-1}(\sin x)$. Without using a calculator, sketch $f(x)$, and then confirm your graph with your calculator.

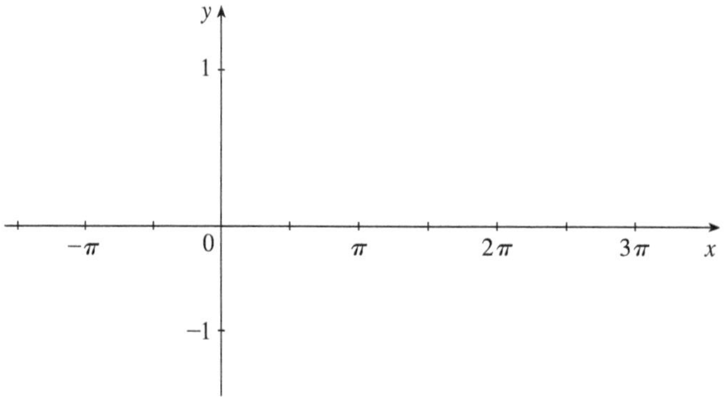

3. Let $g(x) = \sin\left(\sin^{-1} x\right)$. Sketch the graph of g. Is it the same as the graph of f? Why or why not?

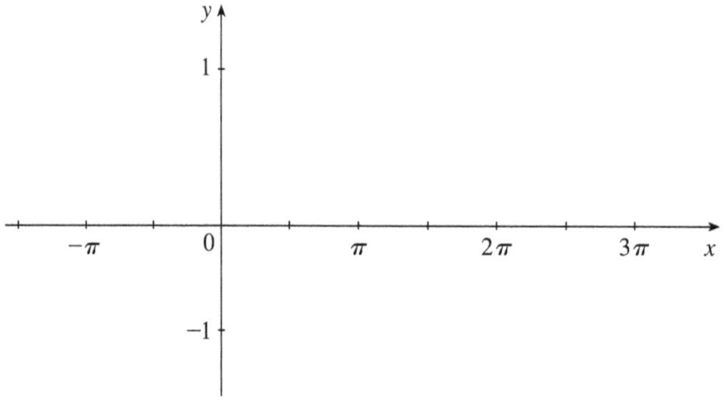

2.6 MODELING HARMONIC MOTION

▼ Suggested Time and Emphasis

1 class. Recommended material.

▼ Points to Stress

1. Understanding the meanings of amplitude, period and frequency, and determining them from the formula $y = a \sin \omega t$ or $y = a \cos \omega t$.
2. Understanding the physical meaning of a phase shift in formulas such as $y = a \sin(\omega(t-c))$ and $y = a \cos(\omega(t-c))$.

▼ Sample Questions

- **Text Question:** What is the relationship between the period of a harmonic function and its frequency?

  Answer: They are reciprocals.

- **Drill Question:** What is the frequency of the harmonic function $y = \sin(6\pi t)$?

  Answer: 3 Hz

▼ In-Class Materials

- An easy way to graph a sine or cosine function given an amplitude and period (no phase shift) is to plot the zeros of one cycle, followed by the maximum and the minimum (between the zeros), and then draw the graph. This method also works with a phase shift, but is not as straightforward.
- Note that not all periodic functions are trigonometric functions. For example, mark a point on a car tire and plot its distance from the ground versus time as the car moves. The resulting graph is called a cycloid and looks like this:

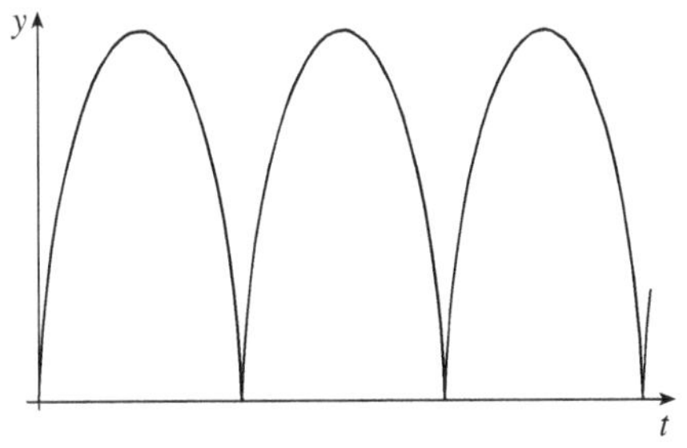

  Terms such as "amplitude", "period", and "frequency" that are used in this section can be used to describe any sort of periodic function.

- Musical notes can be thought of as harmonic motion where the t-axis represents time and the y-axis represents air pressure. A sound wave is a periodic variation of air pressure. The amplitude corresponds to volume, and the frequency corresponds to the note; a higher frequency means a higher pitch. A pure sine or cosine wave sounds like a note played on a tuning fork—not a lot of character. The sound wave from a trumpet looks like this:

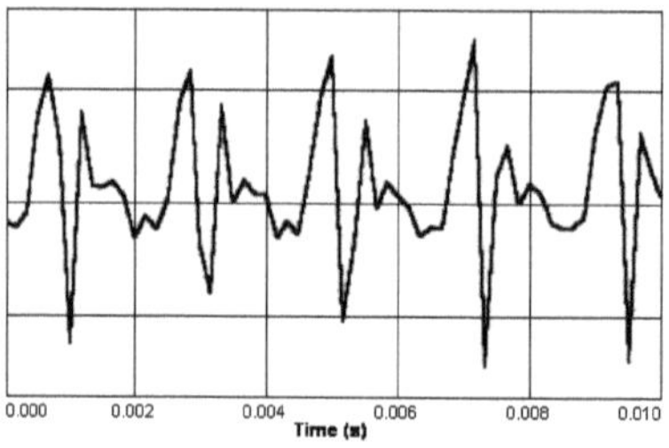

Dept. of Physics, Concordia College, Moorhead MN

- Musical instruments produce a basic sound, plus a series of overtones. We can think of the basic sound as a sine wave with a (relatively) large amplitude, and then the overtones as waves whose frequencies are rational multiples of the basic frequency, and whose amplitudes are smaller. We add the sine curves together to get a sound wave that looks something like this:

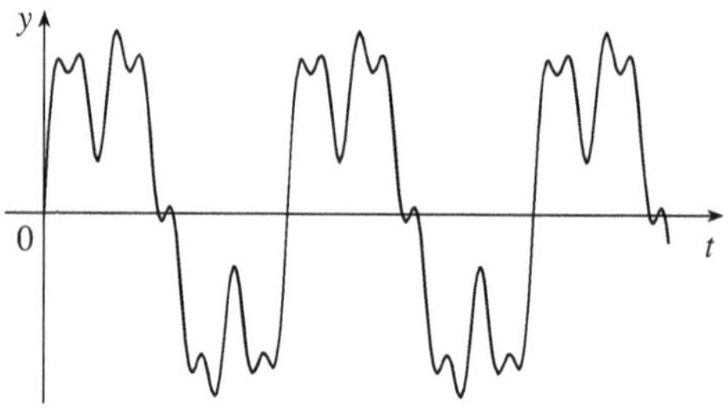

▼ Examples

- $y = 4\sin\left(\pi\left(x - \frac{1}{4}\right)\right)$

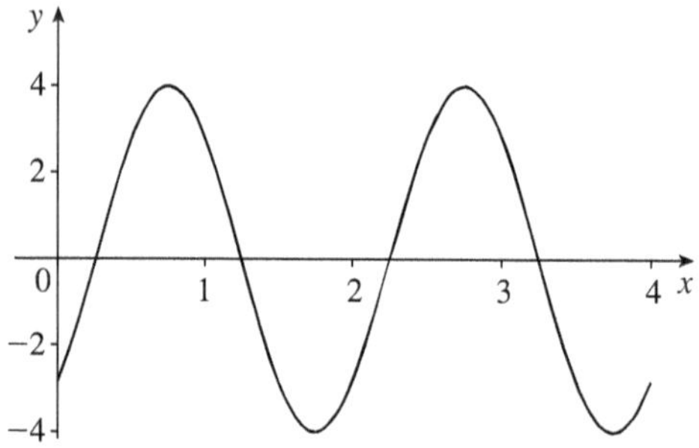

Period: 2, Frequency: $\frac{1}{2}$, Amplitude: 4

• $2\cos\left(\frac{1}{2}x\right)$

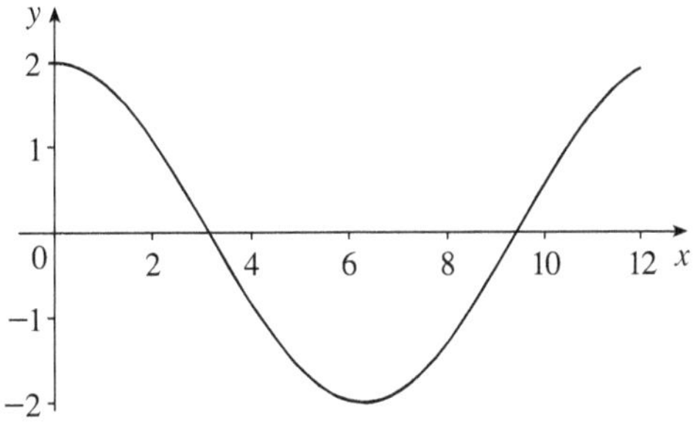

Period: 4π, Frequency: $\frac{1}{4\pi}$, Amplitude: 2

▼ Group Work: The Clock on the Wall

In this activity, students are asked to figure out physical meanings of such quantities as amplitude, frequency, and period.

Answers:

1.

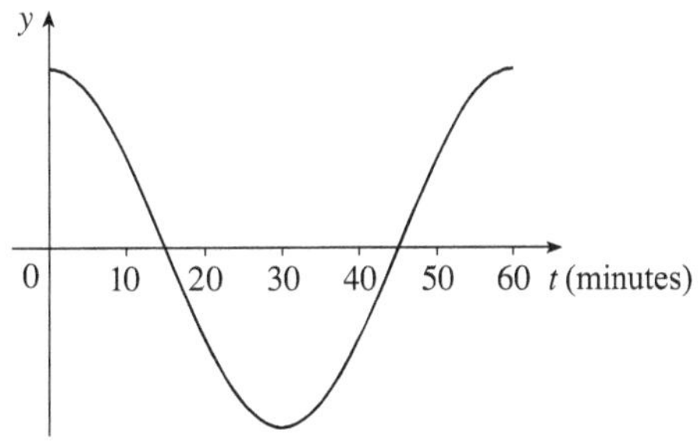

Alternatively, students can measure t in hours, in which $0 \le t \le 1$.

2. Amplitude: Length of the minute hand. Period: 60 minutes (1 hour). Frequency: $\frac{1}{60}$ cycles/minute (or 1 cycle/hour)

3.

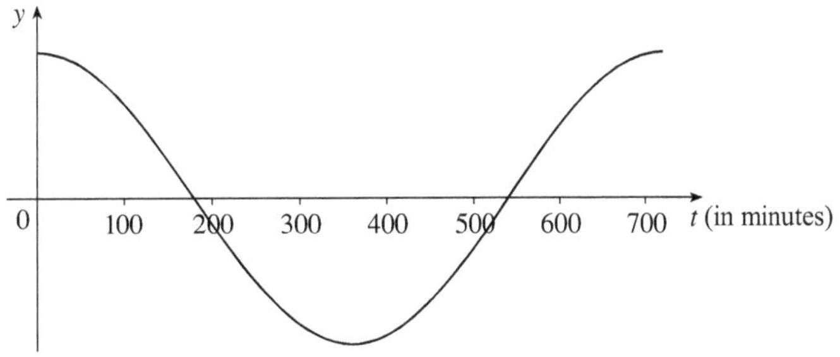

Amplitude: Length of the hour hand. Period: 720 minutes (12 hours). Frequency: $\frac{1}{720}$ cycles/minute (or $\frac{1}{12}$ cycle/hour).

▼ Homework Problems

Core Exercises: 4, 11, 16, 22, 29, 36

Sample Assignment: 4, 9, 11, 14, 16, 17, 22, 26, 29, 33, 36

GROUP WORK, SECTION 2.6

The Clock on the Wall

Consider the clock on the wall of your classroom. Pretend you've drawn a horizontal line between the 3 and the 9 so it looks like this:

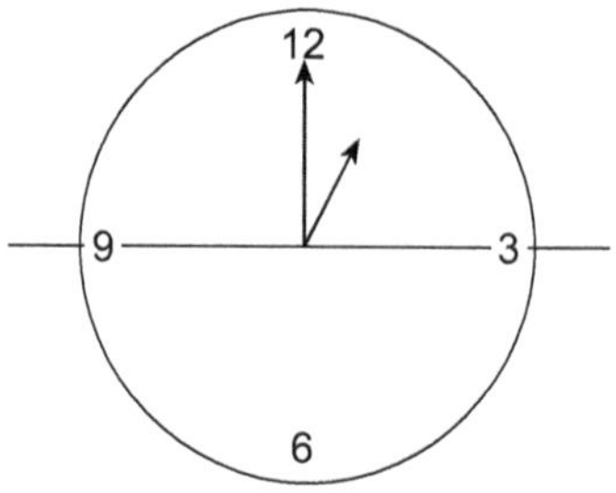

1. Sketch a graph of the height of the tip of the minute hand above this line. Make sure to include units on your graph. When the minute hand is on numbers between 3 and 9, this height will be negative. **Hint:** Your graph will be a trigonometric curve.

2. What is the amplitude of your graph? The period? The frequency?

3. Repeat Exercises 1 and 2 with the hour hand.

3 TRIGONOMETRIC FUNCTIONS: RIGHT TRIANGLE APPROACH

3.1 ANGLE MEASURE

▼ Suggested Time and Emphasis

1 class. Essential material.

▼ Points to Stress

1. Degree and radian measurements of angles.
2. Coterminal angles
3. Arc lengths
4. Linear and Angular speed

▼ Sample Questions

- **Text Question:**

  (a) Which two of the following angles (given in radian measure) are coterminal?

  $0 \qquad \frac{\pi}{2} \qquad \pi \qquad \frac{3\pi}{2} \qquad 2\pi$

  Answer: 0 and 2π

  (b) Which two of the following angles are coterminal?

  $45^\circ \qquad -45^\circ \qquad 180^\circ \qquad -180^\circ \qquad 0^\circ$

  Answer: 180° and -180°

- **Drill Question:** Find an angle with measure (in radians) between $-\frac{\pi}{2}$ and $\frac{\pi}{2}$ that is coterminal with the angle of measure $\frac{11\pi}{3}$ in standard position.

  Answer: $-\frac{\pi}{3}$

▼ In-Class Materials

- Make an analogy between angle measurement and clock measurement. In 48 hours 15 minutes, how far will the minute hand have moved? One answer is that it will go around the clock face $48\frac{1}{4}$ times. Another answer is that it moved one quarter of the way around the clock face, relative to where it started.
- The formula for the length of a circular arc can be verified experimentally. Have students either draw circles or use trash can lids, soup cans, plastic cups, pie-tins, baking pans, or what have you. Measure arc length and radius for 90° angles, 45° angles, and so on, verifying that $s = r\theta$.
- Students often get the false impression that angles with integer measure are always in degrees, and that angles with measures that are rational multiples of π are in radians. Ask students to sketch a 1° angle and then a 1 radian angle. Similarly, ask students to sketch angles with measures $\frac{\pi}{2}$ radians and $\frac{\pi}{2}$ degrees.
- Show that the familiar formulas $A = \pi r^2$ and $C = 2\pi r$ are actually special cases of formulas in this section.
- One can demonstrate the relationship between linear and angular speed this way: The teacher sits in a rotating office chair, and spins at a rate of, say, 2π radians per five seconds. Now have one student stand close to the professor and one farther away, and have them both try to run in circles at that rate. Note that

the one farther away has to travel much faster, demonstrating that if the angular speed is held constant, and r is increased, then v is increased as well.

- If the students have taken some physics, present them with this paradox: We know that Einstein said that the speed of an object cannot exceed 186,000 miles per second, regardless of reference frame. Let the reference frame be a basketball. Have them calculate how fast the sun is moving if the basketball is spinning on a finger, relative to the frame of reference of the basketball. Then have them calculate how fast the stars are moving. These numbers will be faster than the speed of light! This "paradox" is actually a misunderstanding of relativity. The rules about the speed of light being a maximum (and most of relativity theory) apply to what is called an "inertial" reference frame — one where objects in motion remain in motion, and objects at rest remain at rest. The reference frames of a spinning basketball or an accelerating train are not inertial reference frames.

▼ Examples

- We can approximate the shape of the Earth by a sphere of radius 3960 miles. If we wanted to walk far enough to traverse exactly one degree in latitude, how far a trip would we have to take?
  Answer: $3960 \cdot 1 \cdot \frac{\pi}{180} \approx 69$ miles
- Find an angle between $0°$ and $360°$ that is coterminal with $-3624°$.
  Answer: $336°$
- Find an angle between 0 and 2π that is coterminal with $\frac{88\pi}{3}$.
  Answer: $\frac{4\pi}{3}$
- Find an angle between 0 and 2π that is coterminal with 100.
  Answer: $100 - 30\pi \approx 5.7522$

▼ Group Work

There is no group work better than Exercise 74. Ideally, if you are in contact with a teacher at another school (or if your students have willing relatives that live far away) this experiment can be duplicated. We all *believe* that the Earth is big and round, but this experiment allows us to prove it.

Answer: $500 = r\,(7.2)\left(\frac{\pi}{180}\right)$, so $r \approx 3978.9$ mi and $C \approx 25{,}000.0$ mi.

▼ Homework Problems

Core Exercises: 4, 7, 16, 20, 51, 56, 59, 63, 70, 74

Sample Assignment: 4, 7, 10, 16, 20, 23, 30, 34, 51, 53, 56, 59, 63, 68, 70, 74, 86 8

3.2 TRIGONOMETRY OF RIGHT TRIANGLES

▼ Suggested Time and Emphasis

$\frac{1}{2}$–1 class. Essential material.

▼ Points to Stress

1. Definition of the six trigonometric functions as ratios of sides of right triangles.

2. Special triangles: $30°$-$60°$-$90°$ and $45°$-$45°$-$90°$.

3. Applications that involve solving right triangles.

▼ Sample Questions

- **Text Question:** Define "angle of elevation".

  Answers will vary

- **Drill Question:** Consider this triangle:

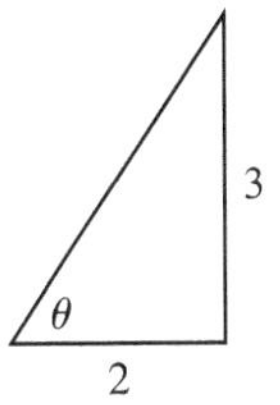

  Find $\sin\theta$, $\sec\theta$, and $\tan\theta$.

  Answer: $\frac{3}{\sqrt{13}}$, $\frac{\sqrt{13}}{2}$, $\frac{3}{2}$

▼ In-Class Materials

Note: If Chapter 2was covered, the first two points may already have been discussed.

- The etymology of the word sine is fairly interesting and not that well known. It starts with the Indian word *jya*, meaning "cord of a bowstring". The Arabs translated this word as *jiba*. The written language had no vowels, so it looked like this: *jb*. In 1145, the Spanish translator Robert of Cheste had to figure out what vowels to put in. Because of the shape of the curve, he thought the word was *jaib*, which meant the opening of a garment that shows a woman's cleavage. So he used the Latin word for the cavity formed by a curve: *sinus*. (This word exists in the English language today — right behind your nose are the sinus cavities.)
- There is an inconsistency in mathematical notation that can be made explicit at this time. We say that $\frac{1}{3} = 3^{-1}$. Similarly, we say that $\frac{1}{\pi} = \pi^{-1}$. But $\dfrac{1}{\sin x}$ is written as $\csc x$. Unfortunately, there is a symbol, $\sin^{-1} x$ that means something entirely different—the arcsine of x. Even worse, as noted in the section, $\sin^2 x = (\sin x)^2$ and $\sin^3 x = (\sin x)^3$. The exponent -1 is a notational anomaly.
- Example 4 is a good outline for an experiment that can be done in real life. Students can estimate the height of their school, their favorite roller coaster, a flagpole, or anything else. If students measure the distance of ten of their paces, they can "pace off" the distance of shadows, which makes it easier to measure them.

- It is relatively easy to make a device to measure angle of elevation. Tape a soda straw to the bottom of a protractor, and tie a string with a weight to the center. When you sight an object through the straw, you can figure out the angle by noticing where the string falls.

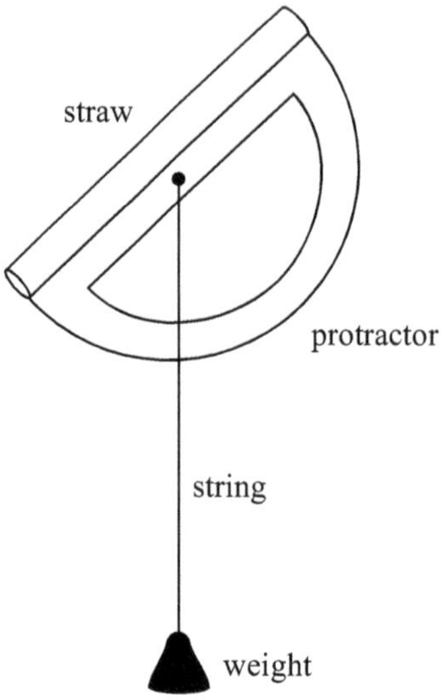

  This device is called a *clinometer*. These days, it is possible to get a clinometer application for your smartphone!

- In addition to figuring out the height of an object, we can also figure out distances. If you know a building is, for example, 700 feet high, you can tell how far away you are by measuring the angle of elevation from where you are standing.

▼ Examples

- Triangles to solve:

1. 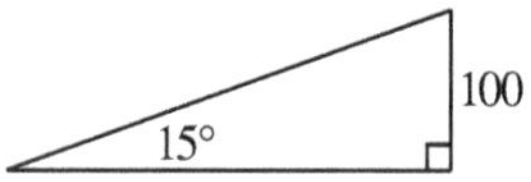

Answer: 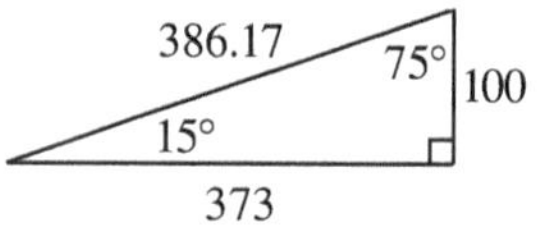

2. 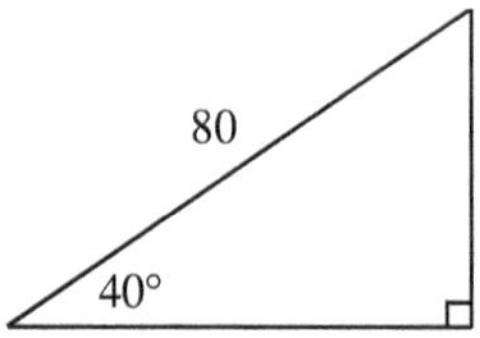

Answer: 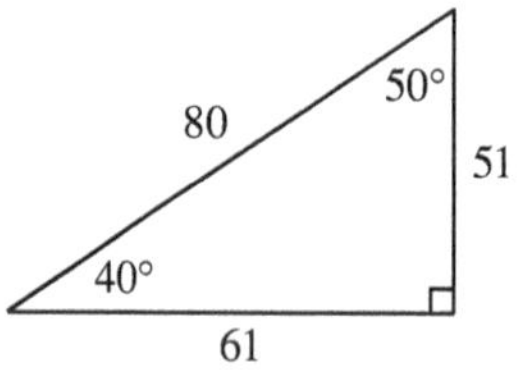

▼ Group Work: Pythagorizing

Students now have several tools to reveal sides and angles of right triangles. They know the trigonometric functions and the Pythagorean Theorem, and they know that the sum of angles in a triangle is always 180°. This activity allows them to assess what they can and can't do at this stage, and to motivate Sections 3.5 and 3.6. It will be interesting to see which students mechanically obtain solutions to Problems 5, 6, and 7, even though they have no tools to work with triangles that are not right triangles. When closing this exercise, make sure to point out that Problem 4 does not have a unique solution.

Answers:

1. $A = 55°$, $b \approx 42.012$, $c \approx 73.246$

2. $A = 55°$, $a \approx 85.689$, $c \approx 104.61$

3. $A = 55°$, $a \approx 49.149$, $b \approx 34.415$

4. The answer is not unique.

5. $A = 55°$, but we do not yet know how to find the other two sides.

6. $A = 55°$, but we do not yet know how to find the other two sides.

7. $A = 55°$, but we do not yet know how to find the other two sides.

8. $A = 90°$, $a \approx 60.926$, $b \approx 10.580$

▼ Homework Problems

Core Exercises: 1, 7, 12, 20, 25, 34, 42, 51, 54

Sample Assignment: 1, 7, 10, 12, 15, 20, 22, 25, 28, 34, 42, 46, 51, 54, 61, 65

GROUP WORK, SECTION 3.2

Pythagorizing

We normally label the angles of a triangle with capital letters, and the corresponding sides with lower case letters:

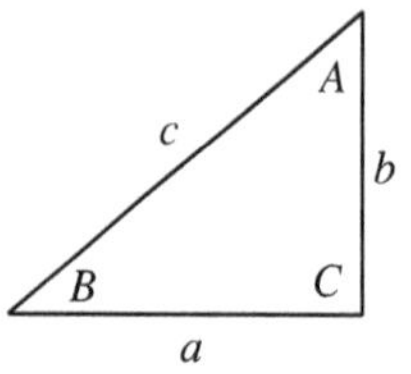

Let's summarize what we know how to do so far. For each of the questions below, solve the triangle or explain why we aren't yet able to solve it:

1. $C = 90°$, $B = 35°$, $a = 60$

2. $C = 90°$, $B = 35°$, $b = 60$

3. $C = 90°$, $B = 35°$, $c = 60$

4. $C = 90°$, $B = 35°$, $A = 55°$

5. $C = 80°$, $B = 35°$, $a = 60$

6. $C = 80°$, $B = 35°$, $b = 60$

7. $C = 80°$, $B = 35°$, $c = 60$

8. $C = 80°$, $B = 10°$, $c = 60$

3.3 TRIGONOMETRIC FUNCTIONS OF ANGLES

▼ Suggested Time and Emphasis

1 class. Essential material.

▼ Points to Stress

1. Finding the reference angle for a given angle.
2. Using the reference angle to evaluate trigonometric functions.
3. The Pythagorean and reciprocal identities of the trigonometric functions.

▼ Sample Questions

- **Text Question:** Is it possible for two different angles to have the same reference angle?
  Answer: Yes
- **Drill Question:** If $\theta = \frac{17\pi}{6}$, find its reference angle $\bar{\theta}$. In what quadrant is θ?
  Answer: $\bar{\theta} = \frac{\pi}{6}$. θ is in the second quadrant.

▼ In-Class Materials

- Students often make the mistake of assuming that the trig functions are linear functions. Since the class will probably involve computing a few trigonometric functions, one can kill two birds with one stone by computing $\sin\left(\frac{\pi}{2}+\frac{\pi}{3}\right)$, $\sin\frac{\pi}{2}$, and $\sin\frac{\pi}{3}$ to demonstrate that, for example, $\sin\left(\frac{\pi}{2}+\frac{\pi}{3}\right) \neq \sin\frac{\pi}{2}+\sin\frac{\pi}{3}$. Return to this point several times; they will be grateful when they don't make that mistake in later courses.
- Don't allow students to simply memorize the mnemonic "All Students Take Calculus" rather than learning the reason for the signs of the trigonometric functions in different quadrants. For example, make sure they are able to articulate why the cosine function is positive in quadrants I and IV and negative in quadrants II and III.

▼ Examples

- $\sin\frac{\pi}{3} = \frac{\sqrt{3}}{2}$
- $\cos\frac{7\pi}{6} = -\frac{\sqrt{3}}{2}$
- $\tan\frac{11\pi}{4} = -1$
- $\sec\frac{17\pi}{3} = 2$
- $\csc\frac{17\pi}{2} = 1$
- $\cot\frac{121\pi}{6} = \sqrt{3}$

▼ Group Work: The Reference Number

This activity allows students to discover a speedy method for finding reference angles of some commonly discussed rational multiples of π by reducing the fractions to lowest terms.

Answers:

1.

θ	Quadrant of θ	$\bar{\theta}$
0	—	0
$\frac{\pi}{3}$	I	$\frac{\pi}{3}$
$\frac{2\pi}{3}$	II	$\frac{\pi}{3}$
$\frac{3\pi}{3}$	—	0
$\frac{4\pi}{3}$	III	$\frac{\pi}{3}$
$\frac{5\pi}{3}$	IV	$\frac{\pi}{3}$
$\frac{6\pi}{3}$	—	0
$\frac{7\pi}{3}$	I	$\frac{\pi}{3}$
$\frac{8\pi}{3}$	II	$\frac{\pi}{3}$
$\frac{9\pi}{3}$	—	0
$\frac{10\pi}{3}$	III	$\frac{\pi}{3}$
$\frac{11\pi}{3}$	IV	$\frac{\pi}{3}$
$\frac{35\pi}{3}$	IV	$\frac{\pi}{3}$
$\frac{1000\pi}{3}$	I	$\frac{\pi}{3}$

2.

θ	Quadrant of θ	$\bar{\theta}$
0	—	0
$\frac{\pi}{6}$	I	$\frac{\pi}{6}$
$\frac{2\pi}{6}$	I	$\frac{\pi}{3}$
$\frac{3\pi}{6}$	—	$\frac{\pi}{2}$
$\frac{4\pi}{6}$	II	$\frac{\pi}{3}$
$\frac{5\pi}{6}$	II	$\frac{\pi}{6}$
$\frac{6\pi}{6}$	—	0
$\frac{7\pi}{6}$	III	$\frac{\pi}{6}$
$\frac{8\pi}{6}$	III	$\frac{\pi}{3}$
$\frac{9\pi}{6}$	—	$\frac{\pi}{2}$
$\frac{10\pi}{6}$	IV	$\frac{\pi}{3}$
$\frac{11\pi}{6}$	IV	$\frac{\pi}{6}$
$\frac{35\pi}{6}$	IV	$\frac{\pi}{6}$
$\frac{1000\pi}{6}$	II	$\frac{\pi}{3}$

▼ Homework Problems

Core Exercises: 5, 8, 14, 26, 35, 50, 48, 61, 67

Sample Assignment: 5, 8, 14, 26, 29, 35, 41, 45, 48, 50, 54, 61, 67, 68, 71

GROUP WORK, SECTION 3.3

The Reference Number

1. For each angle θ below, identify its quadrant and find the reference angle $\bar{\theta}$.

θ	Quadrant of θ	$\bar{\theta}$
0		
$\frac{\pi}{3}$		
$\frac{2\pi}{3}$		
$\frac{3\pi}{3}$		
$\frac{4\pi}{3}$		
$\frac{5\pi}{3}$		
$\frac{6\pi}{3}$		
$\frac{7\pi}{3}$		
$\frac{8\pi}{3}$		
$\frac{9\pi}{3}$		
$\frac{10\pi}{3}$		
$\frac{11\pi}{3}$		
$\frac{35\pi}{3}$		
$\frac{1000\pi}{3}$		

2. For each angle θ below, identify its quadrant and find the reference angle $\bar{\theta}$.

t	Quadrant of θ	$\bar{\theta}$
0		
$\frac{\pi}{6}$		
$\frac{2\pi}{6}$		
$\frac{3\pi}{6}$		
$\frac{4\pi}{6}$		
$\frac{5\pi}{6}$		
$\frac{6\pi}{6}$		
$\frac{7\pi}{6}$		
$\frac{8\pi}{6}$		
$\frac{9\pi}{6}$		
$\frac{10\pi}{6}$		
$\frac{11\pi}{6}$		
$\frac{35\pi}{6}$		
$\frac{1000\pi}{6}$		

3.4 INVERSE TRIGONOMETRIC FUNCTIONS AND TRIANGLES

▼ Suggested Time and Emphasis

1 class. Essential material.

▼ Points to Stress

1. Defining and evaluating the inverse sine, cosine, and tangent functions.

2. Composing trigonometric functions with inverse trigonometric functions.

▼ Sample Questions

- **Text Question:** We can say $\sin^{-1} y = x$ means $y = \sin x$. Given this nice definition, why does the text need to fuss with domains and ranges?
  Answer: For a given y between -1 and 1, there are infinitely many values of x such that $y = \sin x$.
- **Drill Question:** Compute all angles y such that $\sin y = -\frac{1}{2}$ and $0 < y < 2\pi$.
  Answer: $\frac{7\pi}{6}$, $\frac{11\pi}{6}$

▼ In-Class Materials

Note: If Chapter 2 has already been covered, the first two points may have already been addressed.

- It may be helpful to start by reviewing general inverse functions (Section 1.8.7) or inverse trigonometric functions in a real number context (if Section 2.5 has already been covered). If $f(x) = x^3$, what is $f^{-1}(x)$? Note that $f(2) = 8$ and $f^{-1}(8) = 2$. Now move to $f(x) = \sqrt[5]{3x+2} \quad \Rightarrow \quad f^{-1}(x) = \dfrac{x^5 - 2}{3}$. Note that $f(5) \approx 1.762$ and $f^{-1}(1.762) \approx 5$. Now explore $f(x) = x^2$, and point out that even though there is no inverse function, we could make things work out if we were allowed to assume that x was positive. This leads to a natural segue into $f(x) = \sin x$.
- Many students may need a review of notation: $\sin^2 x = (\sin x)^2$, $\sin x^2 = \sin(x^2)$, $(\sin x)^{-1} = \dfrac{1}{\sin x} = \csc x$, $\sin x^{-1} = \sin\left(\dfrac{1}{x}\right)$, but $\sin^{-1} x$ represents the inverse sine of x, the arcsine of x, which is not the same as any of the previous functions.
- Present students with this problem: An airplane is flying at 3000 feet, and is approaching a photographer. The photographer wants to take a picture of it when it is 4000 feet away from her. At what angle should she set the camera so that the plane will be in the shot at the right time?

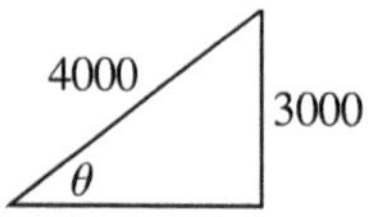

  The answer is $\sin^{-1} \frac{3}{4} \approx 0.848$ radians or $48.59°$. (This problem sets up Group Work 1 in a natural way.)
- In calculus, when integrating certain functions (such as those containing terms of the form $a^2 \pm x^2$) we often obtain unwieldy answers like $\sin(\tan^{-1} x)$. The technique presented in Example 7 and Example 8 (Solution 1) is very useful in that situation.

▼ Examples

- $\sin \frac{\pi}{3} = \frac{\sqrt{3}}{2}$
- $\cos \frac{7\pi}{6} = -\frac{\sqrt{3}}{2}$
- $\tan \frac{11\pi}{4} = -1$
- $\sec \frac{17\pi}{3} = 2$
- $\csc \frac{17\pi}{2} = 1$
- $\cot \frac{121\pi}{6} = \sqrt{3}$

▼ Group Work: The Tripod

Answer: For the first picture: $\theta = \tan^{-1} \frac{150-3}{300} \approx 0.456$ radians, or about $26.1°$. For the second picture, $\theta = \tan^{-1} \frac{100-3}{200} \approx 0.452$ radians, or about $25.9°$. The first picture was taken at a steeper angle.

▼ Homework Problems

Core Exercises: 5, 8, 14, 26, 35, 50, 48, 61, 67

Sample Assignment: 5, 8, 14, 26, 29, 35, 41, 45, 48, 50, 54, 61, 67, 68, 71

GROUP WORK, SECTION 3.4

The Tripod

A photographer is taking pictures of a 150-ft-tall roller coaster. She sets her camera up on a tripod so that the lens of the camera is 3 ft above ground and takes two pictures.

The first picture is taken from 300 feet away with the lens aimed at the top of the roller coaster. The second picture is taken from 200 feet away with the lens aimed two-thirds of the way to the top of the roller coaster.

For which picture is her camera aimed at a steeper angle?

3.5 THE LAW OF SINES

▼ Suggested Time and Emphasis

$\frac{1}{2}$–1 class. Essential material.

▼ Points to Stress

1. Using the Law of Sines to solve triangles.
2. Understanding which cases are ambiguous and which cases are unambiguous.

▼ Sample Questions

- **Text Question:** Why does the book say that SSA is the ambiguous case?

  Answer: It is possible to form two different triangles given two sides and an opposite angle.
- **Drill Question:** Find x in the following diagram:

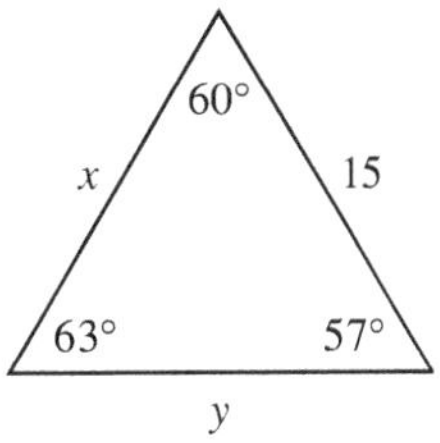

  Answer: $x \approx 14.119$

▼ In-Class Materials

- There is an interesting bit of trivia about the Mount Everest expedition described in the text: Andrew Waugh measured the peak of the mountain to be 29,000 feet exactly. He didn't want people to think his measurement was an estimate (to only two decimal places of accuracy) so he lied and said he measured it to be 29,002 feet.
- The group work for this section asks students to solve SAS and SSS triangles. They won't be able to do this until they learn the Law of Cosines. If you are going to use this group work, then perhaps do not mention which triangles can and cannot be solved with the Law of Sines until after they have tried the group activity.
- One can demonstrate the ambiguous case physically, using two rulers, taped together to make a fixed angle, and a piece of string. See Figure 6 in the text.
- It is tempting to do all examples in degrees. Students should also be exposed to the relatively unfamiliar radian, if possible.

▼ Examples

- ASA
  1. $A = 30°$, $B = 40°$, $c = 100$

     Answer: $A = 30°$, $B = 40°$, $C = 110°$, $a \approx 53.2$, $b \approx 68.4$, $c = 100$
  2. $A = 0.75$ rad, $B = 0.8$ rad, $c = 100$

     Answer: $A = 0.75$ rad, $B = 0.8$ rad, $C \approx \pi - 1.55 \approx 1.59$ rad, $a \approx 68.18$, $b \approx 71.75$, $c = 100$

- SSA. For each of the following, have students draw triangles to try to guess how many solutions there are.
  1. $A = 80°$, $b = 100$, $a = 10$

     Answer: No solution
  2. $A = 80°$, $b = 10$, $a = 100$

     Answer: One solution: $A = 80°$, $B \approx 5.65°$, $C \approx 94.35°$, $a = 100$, $b = 10$, $c \approx 101.25$
  3. $A = 80°$, $b = 121$, $a = 120$

     Answer: Two solutions: $A = 80°$, $B \approx 83.2°$, $C \approx 16.8°$, $a = 120$, $b = 121$, $c \approx 35.2$; $A = 80°$, $B \approx 96.8°$, $C \approx 3.2°$, $a = 120$, $b = 121$, $c \approx 6.9$
  4. $A = 0.2$ rad, $b = 50$, $a = 40$

     Answer: Two solutions: $A = 0.2$ rad, $B \approx 0.25$ rad, $C \approx 2.69$ rad, $a = 40$, $b = 50$, $c \approx 87.8$; $A = 0.2$ rad, $B \approx 2.89$ rad, $C \approx 0.05$ rad, $a = 40$, $b = 50$, $c \approx 10.3$

▼ Group Work: What Can We Do So Far?

Answers:

1. $A = 50°$, $b \approx 83.91$, $c \approx 130.54$
2. No such triangle exists; the angles do not add up to 180°.
3. The answer is not unique.
4. $A = 85°$, $c \approx 86.9$, $b \approx 57.6$
5. $B = 35°$, $c \approx 86.9$, $b \approx 57.6$
6. $A \approx 43.9°$, $B \approx 76.1°$, $b \approx 112.1$
7. We do not yet know how to solve this triangle.
8. We do not yet know how to solve this triangle.
9. We actually can figure this one out. Because $a^2 + b^2 = c^2$, we know that $C = 90°$. Then the definition of sine gives us $A \approx 53.1°$, $B \approx 36.9°$.

▼ Homework Problems

Core Exercises: 2, 3, 15, 20, 25, 30, 35, 40

Sample Assignment: 2, 3, 6, 11, 15, 17, 20, 25, 28, 30, 32, 35, 37, 40, 43

GROUP WORK, SECTION 3.5

What Can We Do So Far?

As usual, we draw and label an arbitrary triangle as follows:

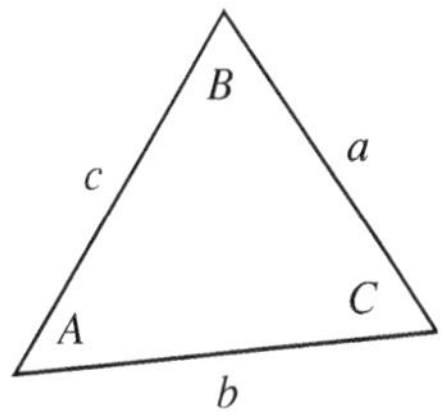

Let's summarize what we know how to do so far. For each of the questions below, solve the triangle or explain why we aren't yet able to solve it.

1. $C = 90°$, $B = 40°$, $a = 100$

2. $C = 90°$, $B = 40°$, $A = 55°$

3. $C = 50°$, $B = 30°$, $A = 100°$

4. $C = 60°$, $B = 35°$, $a = 100$

5. $C = 60°$, $A = 85°$, $a = 100$

6. $C = 60°$, $a = 80$, $c = 100$

7. $C = 60°$, $a = 120$, $b = 90$

8. $a = 40$, $b = 30$, $c = 35$

9. $a = 40$, $b = 30$, $c = 50$

3.6 THE LAW OF COSINES

▼ Suggested Time and Emphasis

$\frac{1}{2}$–1 class. Essential material.

▼ Points to Stress

1. Using the Law of Sines to solve triangles.
2. Using Heron's Formula to find the area of a triangle

▼ Sample Questions

- **Text Question:** Is it possible to find the area of a triangle if the side lengths are known, but the angles are not?

  Answer: Yes
- **Drill Question:** The sides of a triangle are $a = 5$, $b = 8$, and $c = 12$. Find the angle opposite side a.

  Answer: About $18°$ or $\frac{\pi}{10}$ radians.

▼ In-Class Materials

- Have the class apply the Law of Cosines to the special case where C is a right angle. Note that the Law of Cosines is actually a generalization of the Pythagorean theorem.
- Notice that the ability to easily find areas of triangles can be used to help find areas of polygons, because they can be decomposed into triangles.
- If the group work from Section 6.5 was assigned, students can be asked to go back and do the problems that they were not able to do previously.

▼ Examples

- SSS: $a = 4$, $b = 8$, $c = 11$

  Answer: $A \approx 16.2°$, $B \approx 33.9°$, $C \approx 129.8°$
- SAS: $A = 25°$, $b = 10$, $c = 20$

  Answer: $a \approx 11.725$, $B \approx 21.1°$, $C \approx 133.9°$

▼ Group Work: Where Do Good Friends Meet?

Answer: $6\sin(55.77°) = 5\sin(82.81°) \approx 4.96$ ft

▼ Homework Problems

Core Exercises: 2, 4, 7, 16, 19, 22, 27, 41, 44

Sample Assignment: 2, 4, 7, 12, 13, 16, 19, 22, 24, 27, 31, 36, 41, 43, 44, 51

GROUP WORK, SECTION 3.6

Where Do Good Friends Meet?

Every year, the Cedar Falls Chamber of Commerce sponsors a "Chat-a-thon" where happy couples stand four feet apart and compete to see who can spend the most time talking about their relationship. There are two ways that couples tend to lose — they walk away to see what is on television, or they fall asleep standing up. It is actually quite an adorable sight to see two people in love, one six feet tall, the other five feet tall, standing four feet apart, leaning against each other in sleep:

Graphic by Spencer Worobec

How far off the ground are the tops of their heads?

4 ANALYTIC TRIGONOMETRY

4.1 TRIGONOMETRIC IDENTITIES

▼ Suggested Time and Emphasis

$\frac{1}{2}$–1 class. Essential material.

▼ Points to Stress

1. The fundamental trigonometric identities.

2. Proving trigonometric identities.

▼ Sample Questions

- **Text Question:** Given that $\sin^2 x + \cos^2 x = 1$, find an identity for $\tan^2 x + 1$ and show that it is valid.

  Answer: $\dfrac{\sin^2 x}{\cos^2 x} + \dfrac{\cos^2 x}{\cos^2 x} = \dfrac{1}{\cos^2 x} \quad\Rightarrow\quad \tan^2 x + 1 = \sec^2 x$

- **Drill Question:** Prove that $\sin^3 x + \sin x \cos^2 x = \sin x$ is an identity.

  Answer: $\sin^3 x + \sin x \cos^2 x = \sin x \left(\sin^2 x + \cos^2 x\right) = \sin x$

▼ In-Class Materials

- Students often do not grasp the difference between an equation and an identity. Have students classify the following:

$$2x + x = 2x + 2 \qquad \text{(equation)}$$
$$2x + x = 3x \qquad \text{(identity)}$$
$$x + 1 = 2x + 2 \qquad \text{(equation)}$$
$$2(x + 1) = 2x + 2 \qquad \text{(identity)}$$
$$\sin 2x = 2 \sin x \cos x \qquad \text{(identity)}$$
$$2 \sin^2 2x + 2 \cos^2 2x = 2 \qquad \text{(identity)}$$
$$2 \sin^2 2x = 1 \qquad \text{(equation)}$$

- Point out the difference between a reversible operation and an irreversible one. Note the importance—reversible operations can be used to prove identities. Have students classify the following:

$$x^2 \qquad \text{(irreversible)}$$
$$x^3 \qquad \text{(reversible)}$$
$$\sqrt[3]{x} \qquad \text{(reversible)}$$
$$x + c \qquad \text{(reversible)}$$
$$cx \qquad \text{(reversible if } c \neq 0\text{)}$$

▼ Examples

- Proving an identity by combining fractions:

$$\frac{1-\cos\theta}{\sin\theta}-\frac{\sin\theta}{1+\cos\theta}=0$$

Proof:

$$\begin{aligned}\frac{1-\cos\theta}{\sin\theta}-\frac{\sin\theta}{1+\cos\theta} &= \frac{(1+\cos\theta)(1-\cos\theta)}{\sin\theta(1+\cos\theta)}-\frac{\sin^2\theta}{\sin\theta(1+\cos\theta)}\\ &= \frac{(1+\cos\theta)(1-\cos\theta)-\sin^2\theta}{\sin\theta(1+\cos\theta)}\\ &= \frac{\sin^2\theta-\sin^2\theta}{\sin\theta(1+\cos\theta)}=0\end{aligned}$$

- Proving an identity by introducing something extra:

$$\frac{\csc\theta}{\tan\theta+\cot\theta}=\cos\theta$$

Proof:

$$\begin{aligned}\frac{\csc\theta}{\tan\theta+\cot\theta} &= \frac{\csc\theta}{\tan\theta+\cot\theta}\frac{\tan\theta}{\tan\theta}\\ &= \frac{\sec\theta}{\tan^2\theta+1}\\ &= \frac{\sec\theta}{\sec^2\theta}=\cos\theta\end{aligned}$$

▼ Group Work: Finding an Identity

This will allow students to come up with an identity of their own, and touch on the idea of partial fractions.

Answers:

1. $$\begin{aligned}\frac{1}{2b(a-b)}-\frac{1}{2b(a+b)} &= \frac{a+b}{2b(a-b)(a+b)}-\frac{a-b}{2b(a-b)(a+b)}\\ &= \frac{2b}{2b(a^2-b^2)}=\frac{1}{a^2-b^2}\end{aligned}$$

2. Let $a=1$ and $b=\sin\theta$ to obtain $\dfrac{1}{2\sin\theta(1-\sin\theta)}-\dfrac{1}{2\sin\theta(1+\sin\theta)}=\sec^2\theta$.

▼ Homework Problems

Core Exercises: 6, 13, 25, 32, 38, 45, 54, 65, 77, 82, 90, 91

Sample Assignment: 6, 12, 13, 20, 25, 32, 37, 38, 43, 45, 54, 60, 65, 72, 77, 79, 82, 88, 90, 91, 96, 97

GROUP WORK, SECTION 4.1

Finding an Identity

1. Verify the following identity:

$$\frac{1}{a^2 - b^2} = \frac{1}{2b\,(a-b)} - \frac{1}{2b\,(a+b)}$$

2. Use the identity in Problem 1 to find a nice identity whose left-hand side is

$$\frac{1}{2\sin\theta\,(1-\sin\theta)} - \frac{1}{2\sin\theta\,(1+\sin\theta)}$$

4.2 ADDITION AND SUBTRACTION FORMULAS

▼ Suggested Time and Emphasis

$\frac{1}{2}$–1 class. Recommended material.

▼ Points to Stress

1. Addition formulas for sine and cosine.

2. The idea that $a\sin x + b\cos x$ is a phase-shifted sine wave.

▼ Sample Questions

- **Text Question:** The text has an example in which $\cos 75°$ is calculated. How is this done?

  Answer: Write $\cos 75° = \cos(30° + 45°)$ and use an addition formula.
- **Drill Question:** Find $\sin\left(\frac{\pi}{3} + x\right)$ if $\sin x \approx 0.42264$ and $\cos x \approx -0.75984$.

  Answer: $\sin\left(\frac{\pi}{3} + x\right) = \sin\frac{\pi}{3}\cos x + \cos\frac{\pi}{3}\sin x \approx \frac{\sqrt{3}}{2}(-0.75984) + \frac{1}{2}(0.42264) \approx -0.44672$

▼ In-Class Materials

- It might be a good idea for the class to work together to come up with some mnemonics to help them remember the addition formulas. The very act of trying to invent such a mnemonic forces students to look at the formulas carefully, and think about them, which is a memory aid in itself. Students might come up with such things as "the sign of the sine doesn't change" or "the *c*osine *k*eeps like functions together, while the *s*ine *s*witches them".
- Students can test the formulas by seeing if they work for values of sine and cosine that they know. For example, they can look at $\sin(2 - 2)$ and $\cos(\beta - \beta)$ to make sure they get 0 and 1, respectively. They can look at $\sin(60° + 30°)$ and $\cos\left(\frac{\pi}{4} + \frac{\pi}{2}\right)$ to make sure things work out as they should.
- Section 4.3 can be foreshadowed here. Students can be asked to compute $\sin(x + x)$, $\cos(x + x)$, and $\tan(x + x)$. Group Work 1 foreshadows the product-to-sum formulas.
- Graph $\sin x$, $4\cos x$, $2.5\sin x$, and $2.5\cos x$ like so:

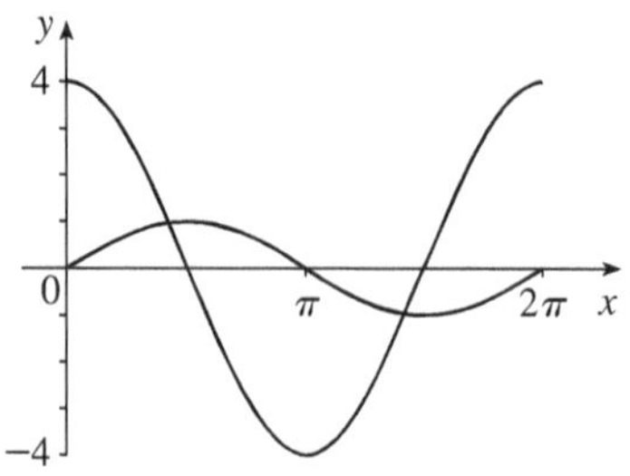

$\sin x$ and $4\cos x$

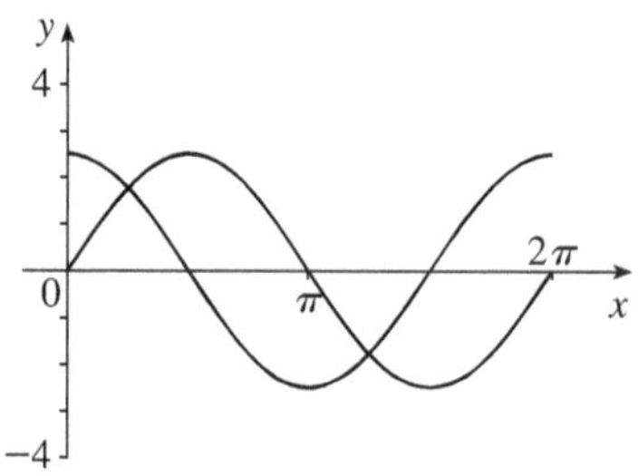

$2.5\sin x$ and $2.5\cos x$

Have students predict which will be bigger, the sum of the first two graphs or the sum of the second two. Note that the sum of the amplitudes of each pair of graphs is 5. Then use the formula from this section to show that the first sum has an amplitude of $\sqrt{17}$ and the second has an amplitude of $\sqrt{12.5}$. Finally, show them the graphs of the sums.

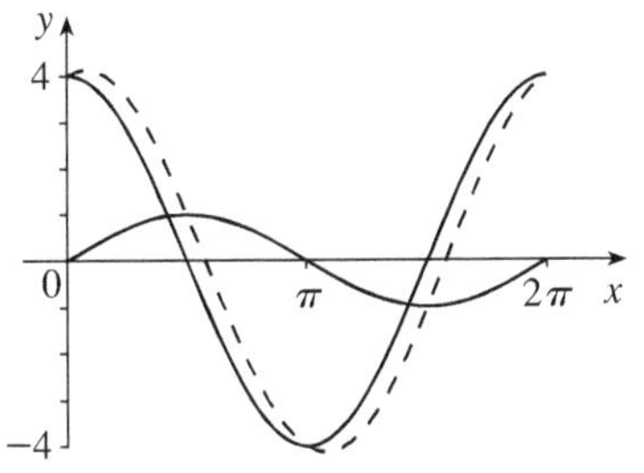

$\sin x + 4\cos x$

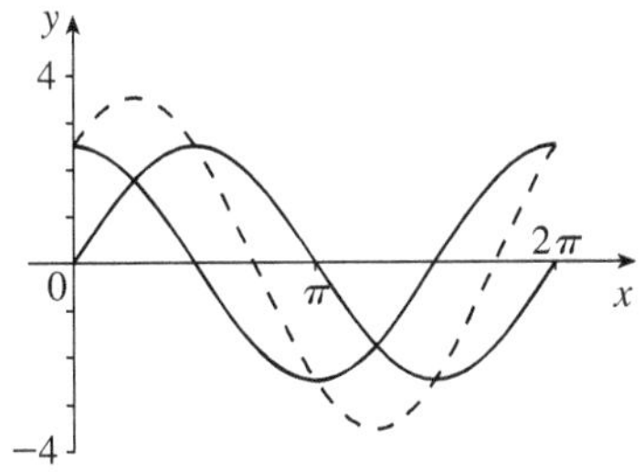

$2.5\sin x + 2.5\cos x$

▼ Examples

- Given that $\sin\frac{\pi}{5} \approx 0.587\,79$, find $\cos\frac{11}{30}\pi$.

  Answer: We verify that $\frac{\pi}{5} + \frac{\pi}{6} = \frac{11}{30}\pi$ and calculate $\cos\frac{11}{30}\pi = \cos\frac{\pi}{5}\cos\frac{\pi}{6} - \sin\frac{\pi}{5}\sin\frac{\pi}{6}$. Since $\frac{\pi}{5}$ is in the first quadrant, we have $\cos\frac{\pi}{5} \approx 0.80902$. Therefore, $\cos\frac{11}{30}\pi \approx 0.40674$.
- Compute $\tan\left(\frac{\pi}{3} + \frac{\pi}{4}\right)$.

  Answer: $\tan\left(\frac{\pi}{3} + \frac{\pi}{4}\right) = -\sqrt{3} - 2$

▼ Group Work: Products to Sums

This allows students to derive the Product-Sum Formulas from Section 4.3.

A good hint for Problem 2 is that we can write $\cos u \sin v$ as $\sin v \cos u$ and then modify the result from Problem 1.

Answers:

1. (a) $\sin(u+v) + \sin(u-v) = 2\sin u \cos v$

(b) $\sin u \cos v = \frac{1}{2}[\sin(u+v) + \sin(u-v)]$

2. $\frac{1}{2}[\sin(v+u) + \sin(v-u)]$ or $\frac{1}{2}[\sin(u+v) - \sin(u-v)]$

3. $\cos u \cos v = \frac{1}{2}[\cos(v+u) + \cos(v-u)]$, $\sin u \sin v = \frac{1}{2}[\cos(u-v) - \cos(u+v)]$

▼ Homework Problems

Core Exercises: 5, 16, 22, 27, 32, 35, 42, 52, 61, 70

Sample Assignment: 5, 12, 16, 19, 22, 27, 31, 32, 35, 38, 42, 43, 48, 52, 56, 61, 70

GROUP WORK, SECTION 4.2

Products to Sums

In this activity, we use the addition and subtraction formulas to find a useful identity for the products of sine and cosine functions.

1. (a) Compute $\sin(u+v)+\sin(u-v)$.

(b) Use your answer from part (a) to find a formula for $\sin u \cos v$.

2. Use your answer from Problem 1 to find an identity for $\cos u \sin v$.

3. Now use a similar technique to the one you used in Problems 1 and 2 to find formulas for $\cos u \cos v$ and $\sin u \sin v$.

4.3 DOUBLE-ANGLE, HALF-ANGLE, AND PRODUCT-SUM FORMULAS

▼ Suggested Time and Emphasis

$\frac{1}{2}$–1 class. Recommended material.

▼ Points to Stress

1. Double-angle formulas for sine and cosine.

2. Power reduction formulas for sine and cosine.

▼ Sample Questions

- **Text Question:** How did we obtain the formula $\sin 2x = 2\sin x \cos x$?

  Answer: We wrote $\sin 2x = \sin(x + x)$ and used an addition formula.

- **Drill Question:** Use product-to-sum formulas to compute $\sin 15° \cos 15°$.

  Answer: $\sin 15° \cos 15° = \frac{1}{2}(\sin 30° + \sin 0°) = \frac{1}{4}$

▼ In-Class Materials

- One fun way to verify the formulas in this section is to compute $\sin x = \sin\left(2 \cdot \frac{1}{2}x\right)$ "the long way". (An even more fun method is to have your students do it.)
- One can illustrate many of the formulas in this section graphically. For example, this is how you can illustrate $\sin^2 x = \dfrac{1 - \cos 2x}{2}$:

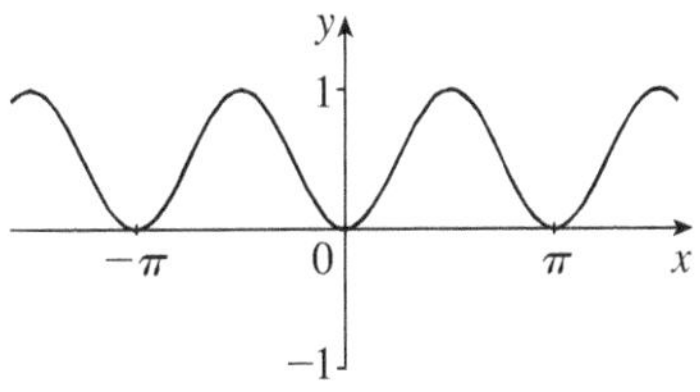

$\sin^2 x$

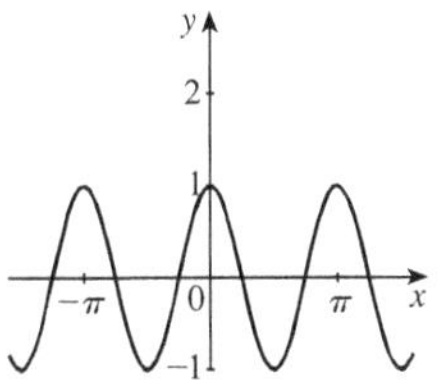

$\cos 2x$

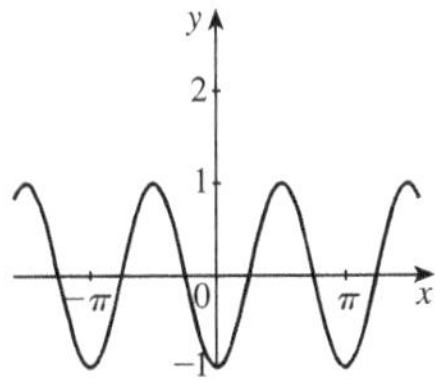

$-\cos 2x$

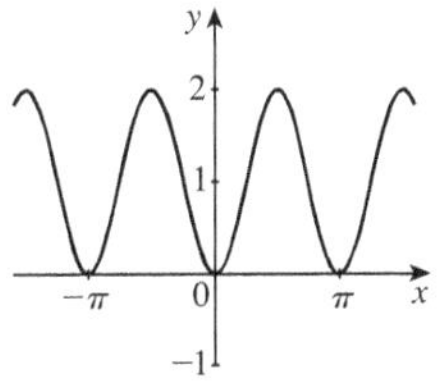

$1 - \cos 2x$

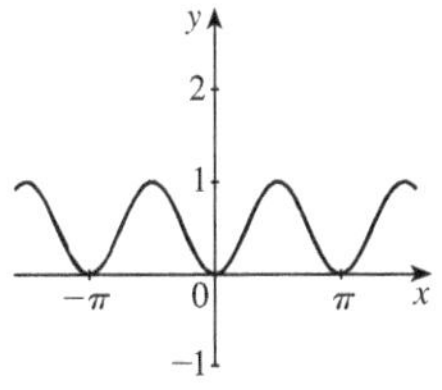

$\dfrac{1 - \cos 2x}{2}$

- Note that we can derive the formula for $\tan\frac{u}{2}$ by dividing the formula for $\sin\frac{u}{2}$ by the formula for $\cos\frac{u}{2}$.

▼ Examples

- $\sin 15° \cos 45° = \frac{1}{2}[\sin 60° + \sin(-30°)] = \dfrac{\sqrt{3} - 1}{4}$
- $\tan^2 22.5° = \dfrac{1 - \cos 45°}{1 + \cos 45°} = \dfrac{2 - \sqrt{2}}{2 + \sqrt{2}}$

▼ Group Work: Pushing Our Luck

This activity gives students practice working with identities, and helps them to understand the derivation of the double angle formulas. Students can check their answers to question 2 by graphing $\cos(3x)$ and then their proposed solution.

Answers:

1. Answers will vary. One correct answer:

$$\begin{aligned}\sin 3x &= \sin(x+2x)\\ &= \sin x\cos 2x + \cos x \sin 2x\\ &= \sin x\left(2\cos^2 x - 1\right) + 2\cos^2 x\sin x\\ &= \sin x\left(2\cos^2 x - 1 + 2\cos^2 x\right)\\ &= \sin x\left(4\cos^2 x - 1\right)\end{aligned}$$

2. Answers will vary. One good answer: $\cos 3x = \cos x\left(1 - 4\sin^2 x\right)$.

▼ Homework Problems

Core Exercises: 3, 12, 20, 31, 38, 52, 59, 75, 84, 103

Sample Assignment: 3, 8, 12, 15, 20, 25, 31, 34, 38, 46, 52, 59, 61, 75, 84, 88, 96, 103

GROUP WORK, SECTION 4.3

Pushing Our Luck

We were able to use the simple facts that $\sin 2x = \sin(x+x)$ and $\cos 2x = \cos(x+x)$ to derive the handy double angle formulas:

$$\begin{aligned} \sin 2x &= 2\sin x\cos x \\ \cos 2x &= \cos^2 x - \sin^2 x \\ &= 1 - 2\sin^2 x \\ &= 2\cos^2 x - 1 \end{aligned}$$

However, it isn't as easy to get nice results for $\sin 3x$ and $\cos 3x$.

1. Show that $\sin 3x = \sin x\left(4\cos^2 x - 1\right)$.

2. Find the cleanest identity you can for $\cos 3x$.

4.4 BASIC TRIGONOMETRIC EQUATIONS

▼ Suggested Time and Emphasis

1–2 classes. Essential material.

▼ Points to Stress

1. Solving linear and quadratic trigonometric equations.
2. Using identities to put trigonometric equations into linear or quadratic form.

▼ Sample Questions

- **Text Question:** In the text, the equation $2\cos^2 x - 7\cos x + 3 = 0$ is solved. The authors wrote it this way: $(2\cos x - 1)(\cos x - 3) = 0$. They then dealt with the first of the two terms, but ignored the second. Why?

  Answer: There is no x such that $\cos x = 3$.
- **Drill Question:** Solve $x\sin 3x + \frac{1}{2}x = 0$.

  Answer: $x = 0$ or $x = \frac{7\pi}{18} + \frac{2}{3}\pi k$

▼ In-Class Materials

- Assume we want to solve $\sin x = \frac{1}{3}$. We want our students to enter $\sin^{-1}\frac{1}{3}$ into a calculator, obtaining 0.33983691, then write down one answer $(0.33983691 + 2\pi k)$, and then *figure out* another $(2.8017557 + 2\pi k)$. It is easy for us to forget that the idea of having to work to find an answer *after* the calculator has spoken is antithetical to everything some students believe about technology. Most of their mathematical experiences with technology involve a problem that ends when the "=" key is pressed, save for (perhaps) checking their answer or interpreting the result. This time, the "=" key signals only the beginning of their thinking. Some students flat-out don't believe that they have to go through this process, and wait for the teacher to show them a shortcut or "the real way." If you can disabuse them of this notion, you will have performed a true service to mathematics.
- When we solve trigonometric equations, we often have to deal with restricted domains. Solve $\sin 3x = 0.15$ with the condition that $5\pi \le x \le 7\pi$.

  Answer: $x \approx 19.000$ or 21.840
- Solve an equation in quadratic form, such as $25\cos^2\theta - 5\cos\theta - 2 = 0$.

  Answer: The left-hand side factors, giving $\cos\theta = \frac{2}{5}$ or $\cos\theta = -\frac{1}{5}$. Thus, $\theta \approx 1.15928 \pm 2k\pi$ or $\theta \approx 1.77215 \pm 2k\pi$.

▼ Examples

- Solve $\cos x\tan x + \frac{1}{2}\tan x = 0$.

  Answer: $x = \pi k$ or $\pm\frac{2\pi}{3} + 2\pi k$
- Solve $\dfrac{\sec 5x}{2} = \cos 5x$.

  Answer: $x = \pm\frac{\pi}{20} + \frac{2}{5}\pi k$

▼ Group Work: Critical Days

In the 1970s, biorhythms were a big deal for some people. There were newspaper columns, special machines in grocery stores and bars, books, etc. These days, biorhythms are often used as an example of a pseudo-science. This activity starts out using techniques from this section, then foreshadows ideas from number theory.

Answers:

1. 78
2. $\frac{7}{3} + 28k$, $\frac{35}{3} + 28k$. The fractions result from the fact that you hit the halfway mark at some point *during* the day.
3. 644 days, or a little over 1 year and 9 months
4. 21,252 days, or a little over 58 years and 2 months

▼ Homework Problems

Core Exercises: 2, 9, 19, 32, 44, 56, 57

Sample Assignment: 2, 4, 6, 9, 13, 19, 21, 28, 32, 38, 42, 44, 49, 56, 57, 59

GROUP WORK, SECTION 4.4

Critical Days

Back in the 1970s, there was a fad called "biorhythms". People believed that their physical, emotional, and intellectual abilities could be modeled by the following equations:

$$\text{Physical} : \sin\frac{2\pi t}{23}$$

$$\text{Emotional} : \sin\frac{2\pi t}{28}$$

$$\text{Intellectual} : \sin\frac{2\pi t}{33}$$

In these equations, t represents the number of days a person has been alive. Biorhythm believers feared "critical days" on which one or more of the functions was equal to 0.

1. How many emotional critical days do people experience in their first three years of life?

2. Find an expression that gives all the days where a person's physical function is equal to $\frac{1}{2}$.

3. A "double critical day" occurs when two functions cross the x-axis on the same day. On what day of your life did you experience your first double critical day?

4. Believers lived in special fear of "triple critical days". On which day of your life will you experience your first triple critical day?

4.5 MORE TRIGONOMETRIC EQUATIONS

▼ Suggested Time and Emphasis

$\frac{1}{2}$–1 class. Recommended material.

▼ Points to Stress

1. Solving trigonometric equations using identities.

2. Finding intersection points.

3. Equations of trigonometric functions with multiple angles.

▼ Sample Questions

- **Text Question:** Why do we need to use an identity to solve $1 + \sin\theta = 2\cos^2\theta$, but not $1 + \sin\theta = 2\sin^2\theta$?

  Answer: The second equation can be put into quadratic form with $\sin\theta$ as the variable. The first one is tricky because it has both sine and cosine in it, so our techniques from the previous section do not work.

- **Drill Question:** Find all solutions to $\sin\frac{\theta}{3} = \frac{\sqrt{3}}{2}$.

  Answer: $\pi + 6\pi k$, $2\pi + 6\pi k$

▼ In-Class Materials

- Before discussing this section, it is good to warm up with some problems from the previous section. The solving techniques of this section all wind up with solving a basic trigonometric equation.
- Finding the intersection of two graphs algebraically is no different than solving a trigonometric equation: We set the formulas for the two graphs equal to each other. The converse is true as well: One way to approximate the solutions to a trigonometric equation is to graph each side of the equation, then use graphical methods to approximate where the two sides are equal.

▼ Examples

- Solve $\sin\theta\cos 2 + \cos\theta\sin 2 = 0.2$.

  Answer: $\sin\theta\cos 2 + \cos\theta\sin 2 = 0.2 \Leftrightarrow \sin(\theta + 2) = 0.2 \Leftrightarrow \theta + 2 \approx 0.2014 + 2\pi k$ or $\theta + 2 \approx 2.9402 + 2\pi k \Leftrightarrow \theta \approx -1.7986 + 2\pi k$ or $\theta \approx 0.9402 + 2\pi k$.

- Solve $\cos\sqrt{\theta} = 0.4$ in the interval $0 \le \theta \le 100$.

  Answer: $\cos\sqrt{\theta} = 0.4 \Leftrightarrow \sqrt{\theta} \approx 1.1592 + 2\pi k$ or $\sqrt{\theta} \approx -1.1592 + 2\pi k \Leftrightarrow$ $\sqrt{\theta} \approx 1.1592, 5.1240, 7.4424, 11.4072\ldots$.

  Solutions in the desired interval are $\theta \approx 1.3437$, $\theta \approx 26.2553$, and $\theta \approx 55.3893$.

▼ Group Work 1: Some Unusual Equations

Answers:

1. $x = \frac{\pi}{2} + \pi k$ or equivalently $x = \pm\frac{\pi}{2} + 2\pi k$

2. $x \approx \pm 0.25493168 + \frac{\pi}{2}k$

▼ Group Work 2: A Different Periodic Function

Some students get the impression that the six trig functions are the *only* periodic functions. This activity introduces another periodic function.

Answers:

1.

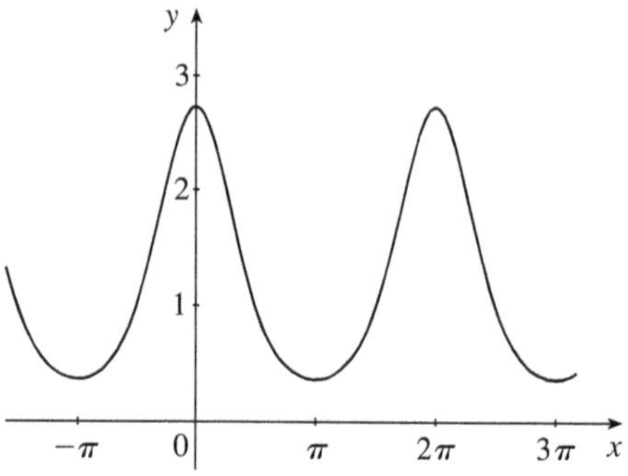

2. $x \approx 1.3457582 + 2\pi k$

3. There is no solution. The graph does not touch the line $y = 2.8$.

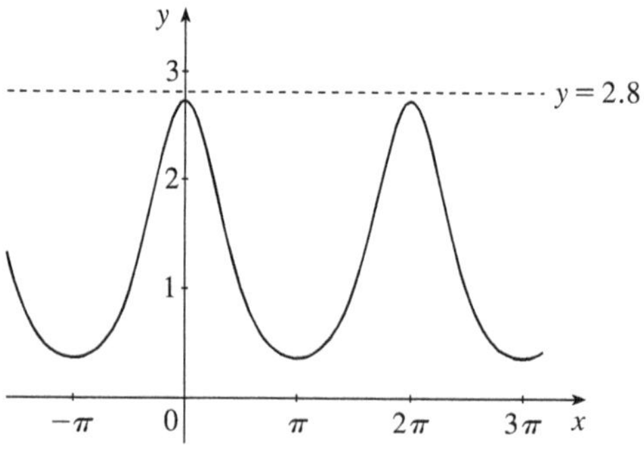

▼ Homework Problems

Core Exercises: 4, 10, 19, 32, 36, 41, 44, 52, 58, 64

Sample Assignment: 4, 10, 15, 19, 24, 32, 34, 36, 39, 41, 44, 48, 52, 55, 58, 62, 64, 65

GROUP WORK 1, SECTION 4.5

Some Unusual Equations

1. Solve $\sin(\cos x) = 0$.

2. Solve $\sin(\cos 4x) = \frac{1}{2}$.

GROUP WORK 2, SECTION 4.5

A Different Periodic Function

1. Graph $e^{\cos x}$.

2. Solve $e^{\cos x} = 1.25$.

3. Solve $e^{\cos x} = 2.8$. How is your solution reflected on your graph?

5 POLAR COORDINATES AND PARAMETRIC EQUATIONS

5.1 POLAR COORDINATES

▼ Suggested Time and Emphasis

1 class. Essential material.

▼ Points to Stress

1. The relationships between rectangular and polar coordinates.
2. Converting equations into different coordinate systems.

▼ Sample Questions

- **Text Question:** Where do the conversion equations $x = r\cos\theta$, $y = r\sin\theta$ come from?

  Answer: Several correct answers are possible. Anything addressing the definitions of sine and cosine, for example, should be given credit.
- **Drill Question:** Convert the equation $r\sin\theta = 3r\cos\theta + 2$ to polar form.

  Answer: $y = 3x + 2$

▼ In-Class Materials

- Begin with an intuitive definition of polar coordinates and then derive the algebraic formulas, noting that the graph of a polar function need not pass the Vertical Line Test.
- Do several examples of converting Cartesian equations into polar equations, such as $y^2 = 4x$ to $r = 4\csc\theta\cot\theta$, and of converting polar equations into Cartesian equations (the result of which are sometimes implicit equations) such as $r = 2\sec\theta$ to $x = 2$ and $r = 2(1 + \cos\theta)$ to $\left(x^2 + y^2 - 2x\right)^2 = 4\left(x^2 + y^2\right)$.
- Point out how some equations are simpler to consider in rectangular coordinates [$y = \ln x$ is easier than $r\sin\theta = \ln r + \ln\cos\theta$] but some equations are simpler in polar coordinates ($r = \theta$, the simple spiral, is much easier than $\pm\sqrt{x^2 + y^2} = \tan^{-1}(y/x)$). You can foreshadow Chapter 7 at this point, pointing out that there are curves called rotated ellipses, hyperbolas and parabolas, that turn out to be very nice when considered as polar equations.

▼ Examples

- Coordinate conversion:

  Rectangular $(8, 16)$ is the same as polar $\left(8\sqrt{5}, 1.107\right)$.

  Rectangular $\left(5, -5\sqrt{3}\right)$ is the same as polar $\left(10, -\frac{\pi}{3}\right)$.

  Polar $\left(4, \frac{13\pi}{6}\right)$ is the same as rectangular $\left(2\sqrt{3}, 2\right)$.

  Polar $(7, 10)$ is (approximately) the same as rectangular $(-5.8735, -3.8081)$.
- Equation conversion:

  Rectangular: $x^2 + y^2 = 9$ is the same as polar $r = 9$.

  Polar $\tan\theta = 1$ is the same as rectangular $y = x$.

▼ Group Work 1: Uniqueness of Representation

We say that one point can have several polar representations, and only one rectangular representation. This activity is designed to solidify this fact, while giving students a chance to practice converting points between coordinate systems. If students are thinking about polar coordinates entirely algebraically, they may have difficulty with Problem 1(d), because $\arctan\left(\frac{-5}{0}\right)$ is undefined. Perhaps they will need the hint that this is not a hard problem if someone just draws a picture. Problem 4 is impossible. The main point of this activity is that it can't be done. Make sure, when closing, that this is stated clearly.

Answers:

1. (a) $\left(\frac{5}{2}\sqrt{2}, \frac{5}{2}\sqrt{2}\right)$ **(b)** $\left(\sqrt{13}, -\arctan\frac{2}{3}\right) \approx \left(\sqrt{13}, -0.588\right)$ **(c)** $\left(4, -\frac{\pi}{3}\right)$ **(d)** $\left(5, \frac{3\pi}{2}\right)$

2. All are the point $\left(-\frac{\sqrt{2}}{2}, -\frac{\sqrt{2}}{2}\right)$.

3. Answers will vary.

4. This is impossible.

▼ Group Work 2: Some Basic Polar Graphs

This will foreshadow the next section, where polar graphs are discussed in earnest.

Answers:

1. The graph is a circle.

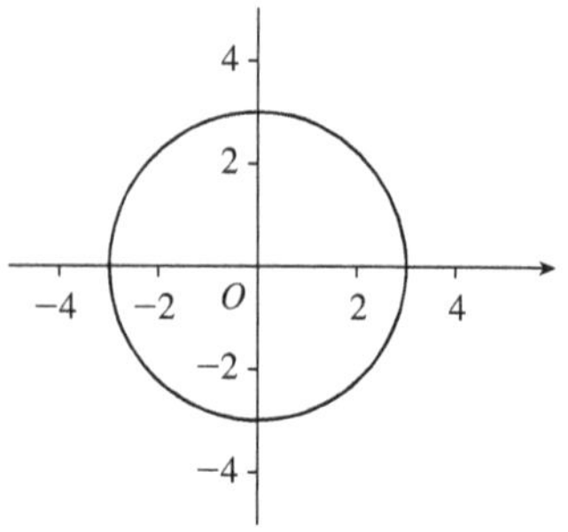

2. The graph is a straight line through the origin.

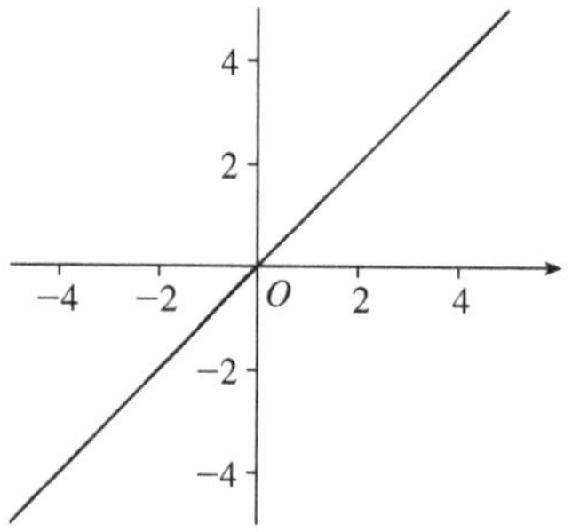

3. The graph is a spiral

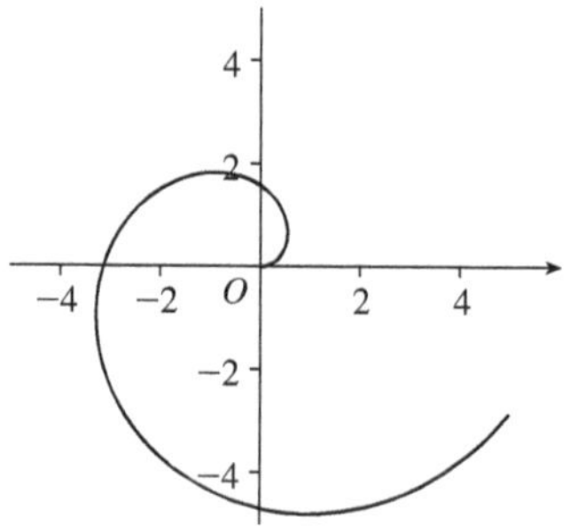

▼ Homework Problems

Core Exercises: 2, 3, 10, 17, 30, 36, 44, 49, 54, 63

Sample Assignment: 2, 3, 6, 10, 17, 22, 23, 26, 30, 33, 36, 40, 44, 47, 49, 54, 63, 68

GROUP WORK 1, SECTION 5.1

Uniqueness of Representation

1. Convert these points (given in rectangular coordinates) to polar coordinates:

(a) $(5, 5)$ **(b)** $(-3, 2)$ **(c)** $(2, -2\sqrt{3})$ **(d)** $(0, -5)$

2. Convert these points (given in polar coordinates) to rectangular coordinates:

(a) $\left(2, \frac{5\pi}{4}\right)$ **(b)** $\left(-2, \frac{\pi}{4}\right)$ **(c)** $\left(2, \frac{13\pi}{4}\right)$ **(d)** $\left(2, -\frac{3\pi}{4}\right)$

3. Choose a point in the first quadrant, and give three different polar representations for that point.

4. Choose a point in the first quadrant, and give three different rectangular representations for that point.

GROUP WORK 2, SECTION 5.1

Some Basic Polar Graphs

Do you remember when you first started drawing graphs? No doubt, at some point, you were given a simple formula like $y = 2x$, and were instructed to plot a bunch of points. You did so and, by Neptune's Knickerbockers, you obtained a line!

In this activity, we are going to draw some polar graphs by plotting a bunch of polar points and seeing what happens.

1. $r = 3$. Plot many points whose first polar coordinate is 3. You will probably start with $(3, 0)$, then maybe $\left(3, \frac{\pi}{6}\right)$, or whatever you like. Choose many values for θ, but make sure to keep $r = 3$. What do you get?

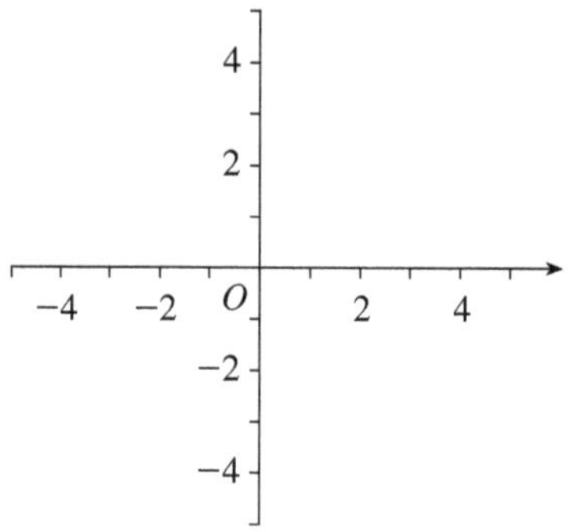

2. $\theta = \frac{5\pi}{4}$. Again, plot some points to figure out what this graph will look like. Your points can have any r-coordinate you chose, but the θ-coordinate must be $\frac{5\pi}{4}$. Make sure to include some negative values of r along with positive ones.

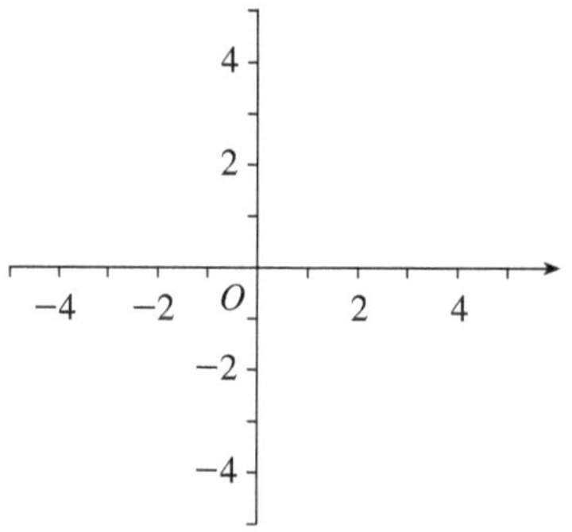

3. $r = \theta, \theta \geq 0$. In this plot, both polar coordinates will always be the same. You will probably want to plot $(0, 0)$ and $\left(\frac{\pi}{2}, \frac{\pi}{2}\right)$ and many others until you understand the shape of this curve.

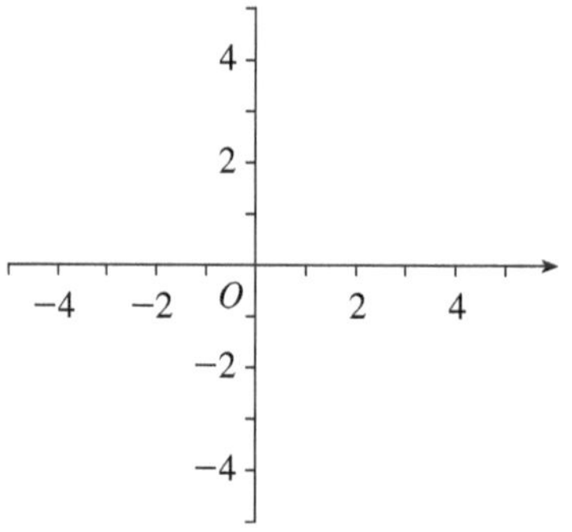

5.2 GRAPHS OF POLAR EQUATIONS

▼ Suggested Time and Emphasis

1 class. Essential material.

▼ Points to Stress

1. Graphs in polar coordinates.
2. Tests for symmetry.
3. Using graphing devices to obtain polar graphs.

▼ Sample Questions

- **Text Question:** Sketch the graph of $r = \cos\theta$.

  Answer:

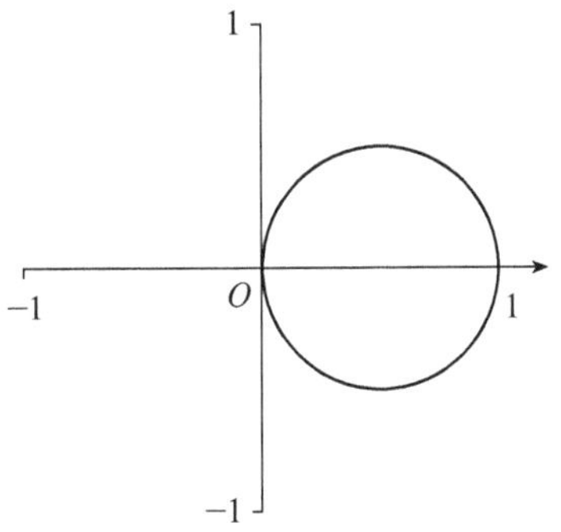

- **Drill Question:** What is the polar equation of a circle with radius 3 centered at the origin?

  Answer: $r = 3$

▼ In-Class Materials

- After showing students how to graph polar functions on a calculator, give them a chance to experiment and try to come up with interesting-looking polar graphs. Make sure they know that if their graphs appear to have a lot of cusps, it could be that their Δt is set too large.
- An interesting graph to look at is $r = \dfrac{\theta}{\theta + 1}$, $\theta \geq 0$.

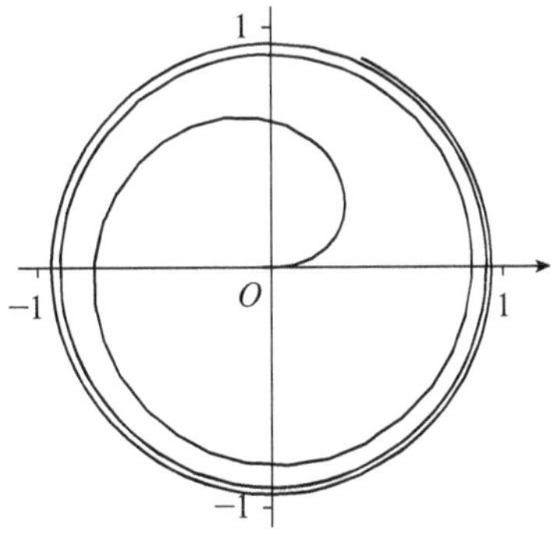

Have the class try to figure out if, as $\theta \to \infty$, the curve will get infinitely close to the curve $r = 1$, and if so, why. Ask if the curve will ever touch $r = 1$. If this curve is combined with the circle $r = 1$, the resulting set of points is what topologists call "connected, but not path-connected". The set of points is not path-connected because there is no path from the origin that touches the outer circle. It is called "connected" because (to simplify things somewhat) there is no curve that separates the two components without touching either. Dr. Shaw calls this curve "The Topologist's Hypno-Disk"

▼ Examples

- Do a problem where $r = f(\theta)$ is graphed both as a function in rectangular coordinates and as a polar function. For example, if $r = (1.5)^{\theta}$ we get the two graphs shown below.

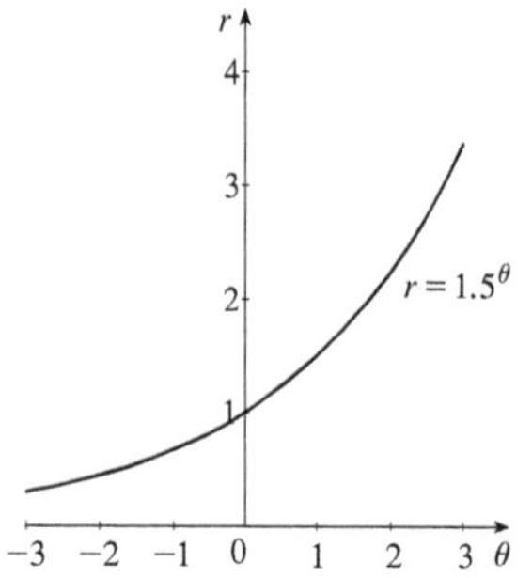

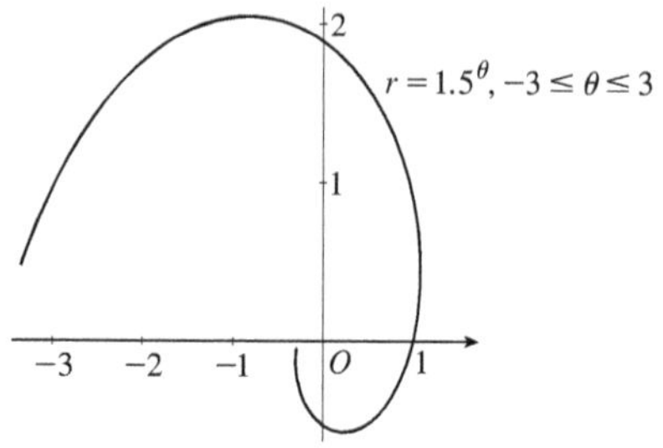

- Sketch a graph of the polar curve $r = f(\theta)$ where $f(\theta)$ is the function whose representation in rectangular coordinates is given below.

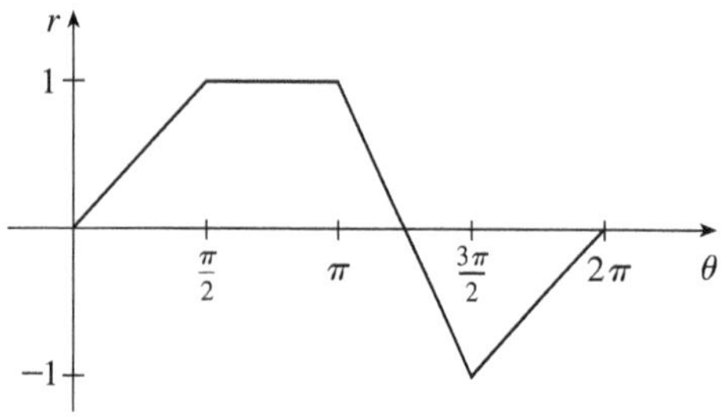

Answer:

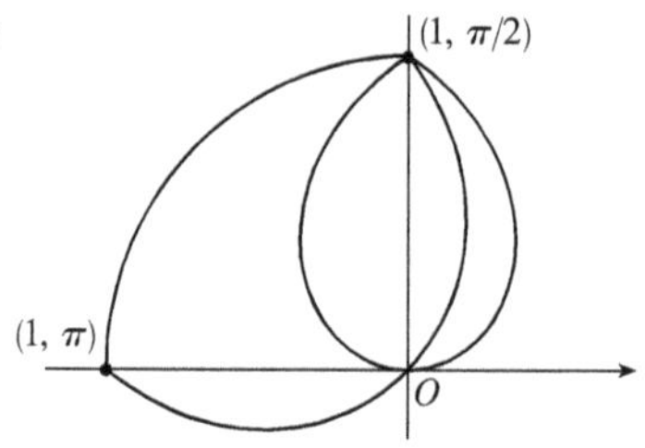

▼ Group Work 1: Picture Pages

This group work is designed to assist students in visualizing a polar function $r = f(\theta)$ graphed in rectangular coordinates.

Answers:

1.

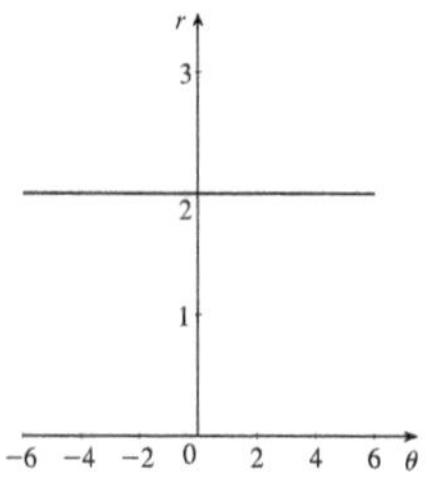

E.

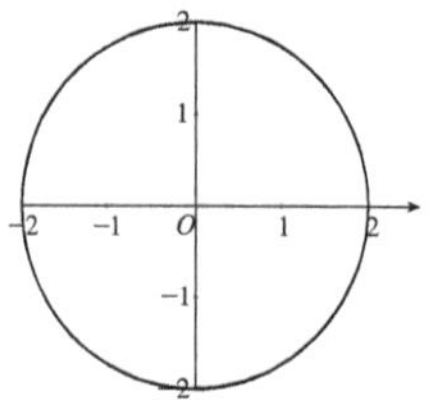

$r = 2$

2.

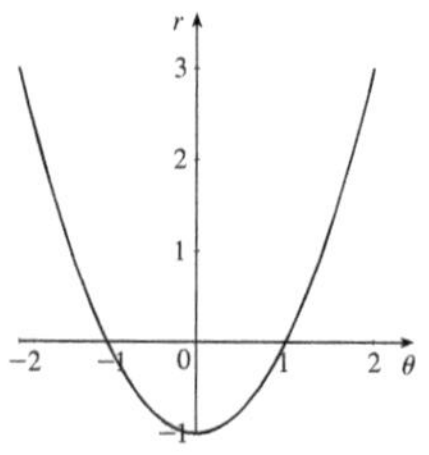

A.

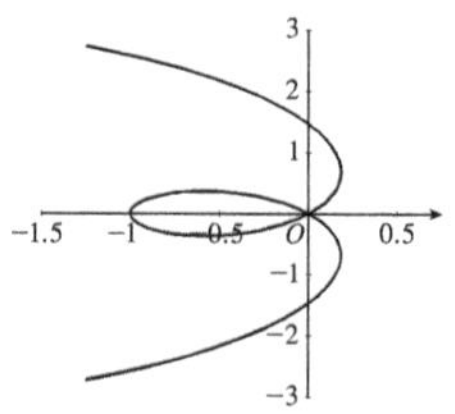

$r = \theta^2 - 1, \theta \in [-2, 2]$

3.

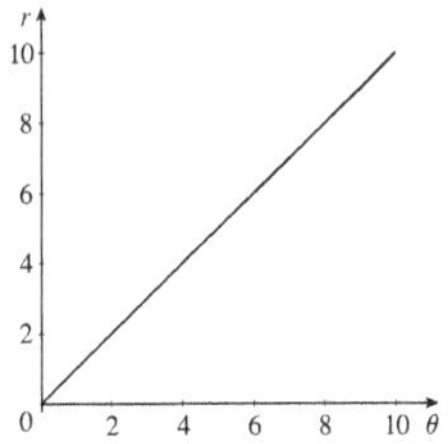

B.

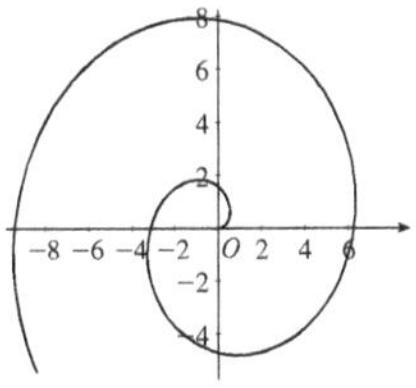

$r = \theta,\ \theta \in [-10, 10]$

4.

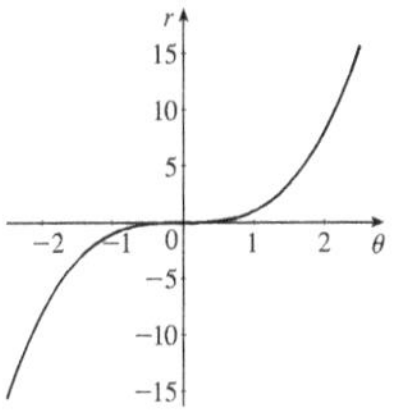

D.

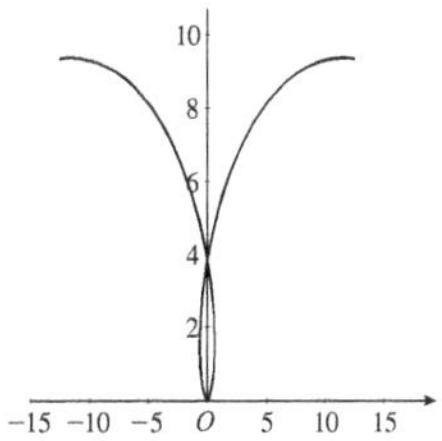

$r = \theta^3,\ \theta \in [2.5, 2.5]$

5.

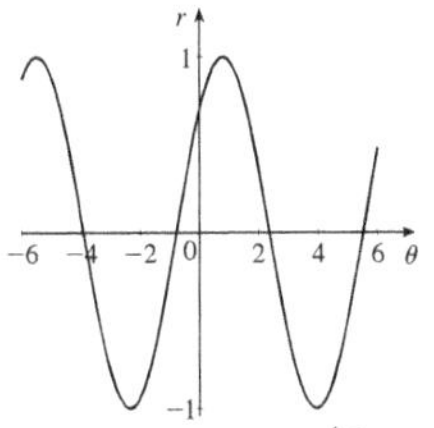

C.

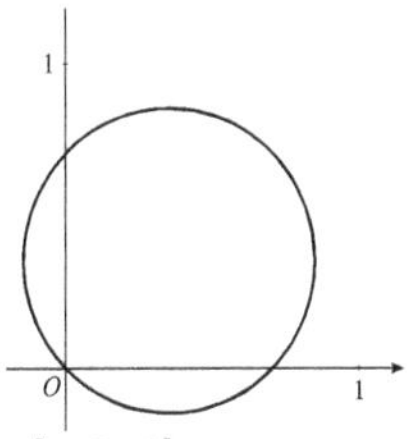

$r = \cos\left(\theta - \frac{\pi}{4}\right),\ \theta \in [-6, 6]$

▼ Group Work 2: Cardioids

Answers:

1.

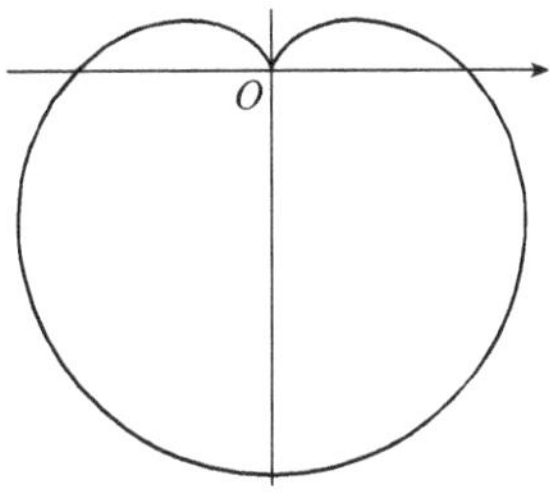

2. $\left(\frac{3}{2}, -\frac{\pi}{6}\right), \left(\frac{3}{2}, -\frac{5}{6}\pi\right)$

3. $\left(2, -\frac{\pi}{2}\right), \left(\frac{1}{2}, \frac{\pi}{6}\right), \left(\frac{1}{2}, \frac{5\pi}{6}\right)$. Note that the derivative is not defined at $\left(0, \frac{\pi}{2}\right)$.

▼ Homework Problems

Core Exercises: 3, 10, 24, 29, 36, 47, 50, 51, 55

Sample Assignment: 3, 8, 10, 13, 19, 24, 29, 36, 42, 47, 49, 50, 51, 52, 55, 57

GROUP WORK 1, SECTION 5.2

Picture Pages

The graphs in the first column show how r varies with θ in rectangular form. The second column shows the polar graphs of these functions. Match up the graphs so that each pair corresponds to one equation for r as a function of θ.

1.

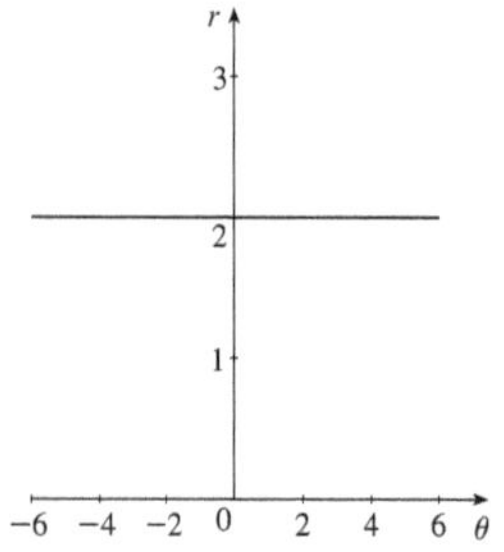

A.

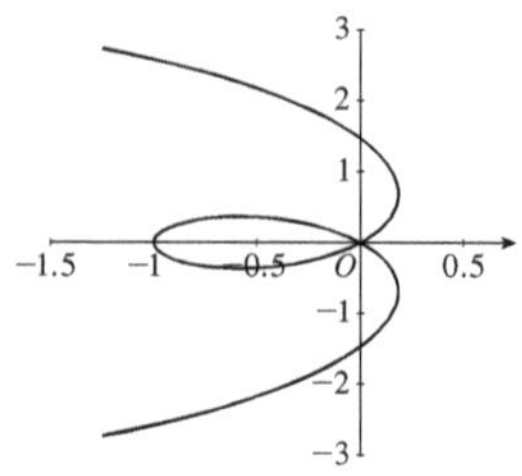

2.

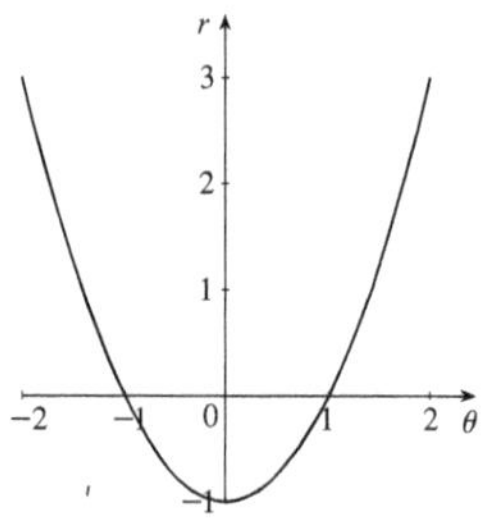

B.

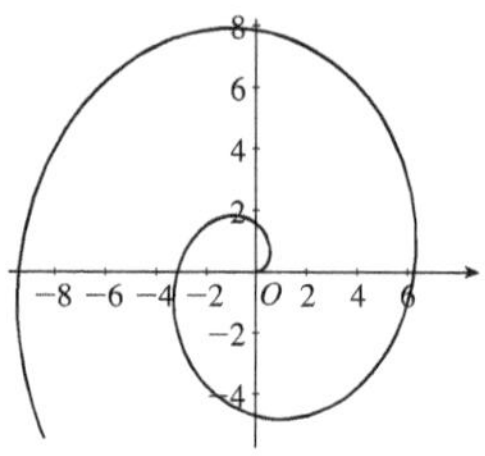

3.

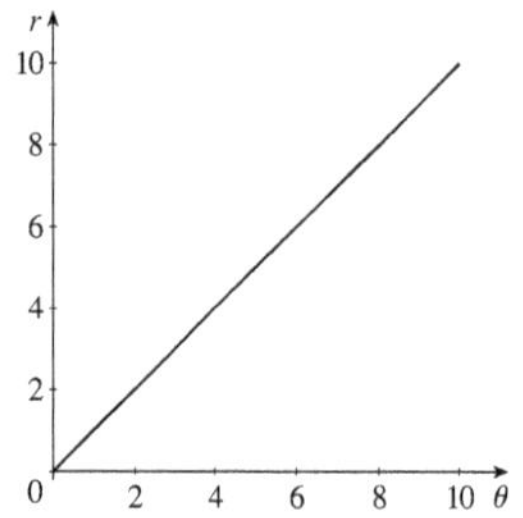

C.

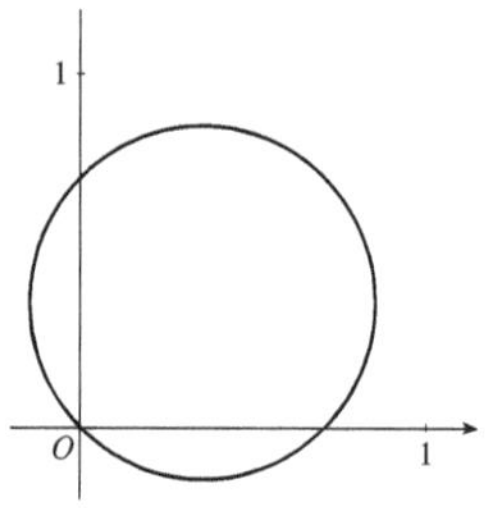

4.

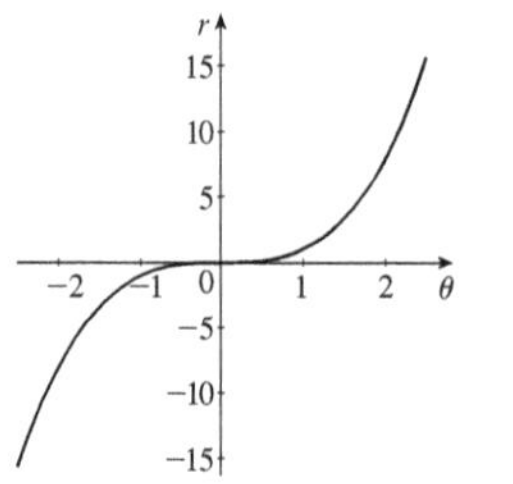

D.

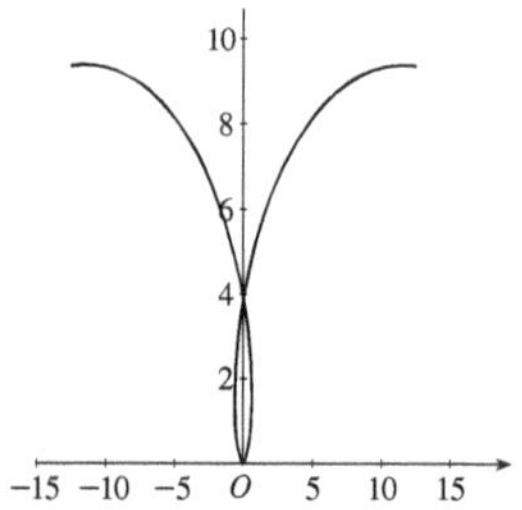

5.

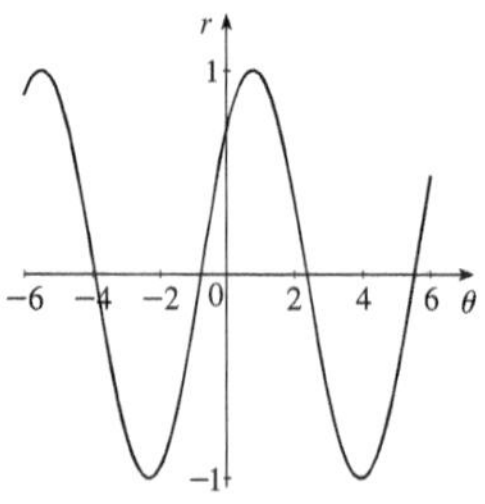

E.

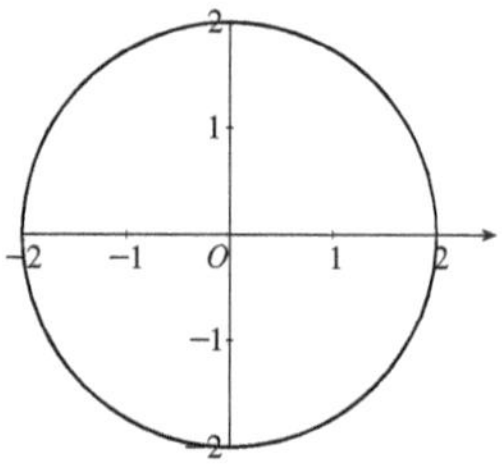

GROUP WORK 2, SECTION 5.2
Cardioids

Consider the polar curve $r = 1 - \sin\theta$

1. Carefully sketch a graph of this curve on the polar graph paper provided, by computing at least 10 points on the curve and connecting the dots.
2. Find polar coordinates for the points on the curve where the tangent line is *vertical.* Find exact values if you can, but if this is impossible, a good estimate will do.

3. Find polar coordinates for the points where the tangent line is *horizontal.* Again, find exact values where possible.

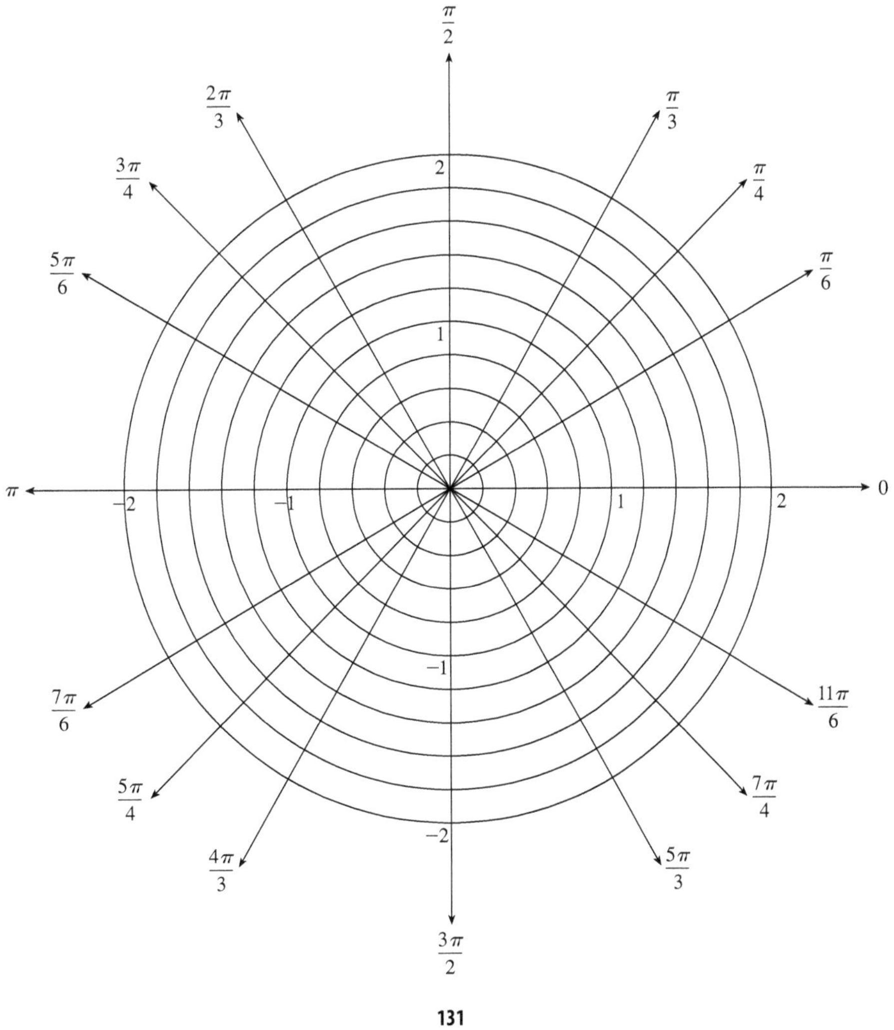

5.3 POLAR FORM OF COMPLEX NUMBERS; DE MOIVRE'S THEOREM

▼ Suggested Time and Emphasis

$\frac{1}{2}$–1 class. Recommended material.

▼ Points to Stress

1. Three representations of a complex number: as a point, as a number $a + bi$, and in trigonometric form.
2. Multiplication and division of complex numbers in trigonometric form.
3. DeMoivre's Theorem

▼ Sample Questions

- **Text Question:** Graph the following set of complex numbers: $\{z \mid |z| = 1\}$

  Answer:

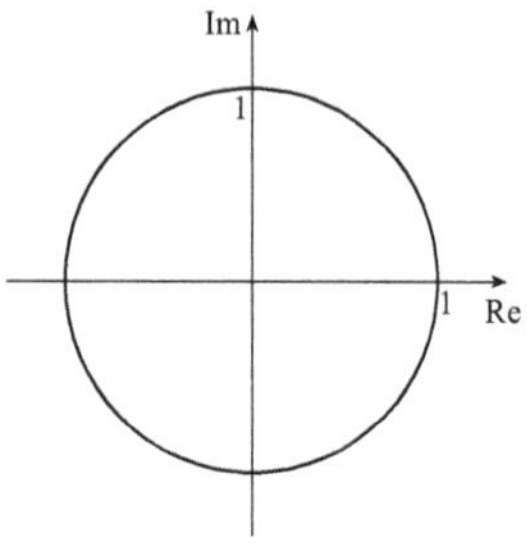

- **Drill Question:** Compute $\left(\sqrt{3} - i\right)^{10}$.

  Answer: $512 + 512\sqrt{3}i$

▼ In-Class Materials

- Notice that we now have three distinct ways of representing the same quantity. Each has advantages and disadvantages. For example, it is cumbersome to compute

$$3\left(\cos\tfrac{\pi}{2} + i\sin\tfrac{\pi}{2}\right) + 2\left(\cos\tfrac{\pi}{6} + i\sin\tfrac{\pi}{6}\right) = \sqrt{19}\left(\cos 1.162 + i\sin 1.162\right)$$

  but not too hard to compute

$$3i + \left(\sqrt{3} + i\right) = \sqrt{3} + 4i$$

  Similarly, it is annoying to divide

$$\frac{3 - 6i}{4 + 2i} = -\tfrac{3}{2}i$$

  but it is not hard to divide

$$\frac{8\left(\cos 4 + i\sin 4\right)}{4\left[\cos(-1) + i\sin(-1)\right]} = 2\left(\cos 5 + i\sin 5\right)$$

- Verify, using the multiplication of complex numbers in trigonometric form, that $i^2 = -1$. This leads to another way to view complex numbers, one that is more intuitively justifiable for some than allowing negative square roots. We define a "complex number" as a point in the plane, where we want to define addition in a certain way (which will parallel vector addition defined in the next section) and multiplication of quantities of the same magnitude to correspond to a rotation. The idea that the point $(0, 1)$ times itself is $(-1, 0)$ is then a logical consequence of this definition of multiplication. Then, if we define the point $(1, 0)$

to be "1" and the point $(0, 1)$ to be "i" we obtain $i^2 = -1$. This is not violating the laws of conventional mathematics; it is simply a result of our definitions of operations on these point quantities. In the real world, it turns out there are quantities (such as waveforms, impedance of an AC circuit, etc.) that are most usefully represented in this way.

- Ask the class, "What are the square roots of 1?" Elicit the answers ± 1, and draw them this way:

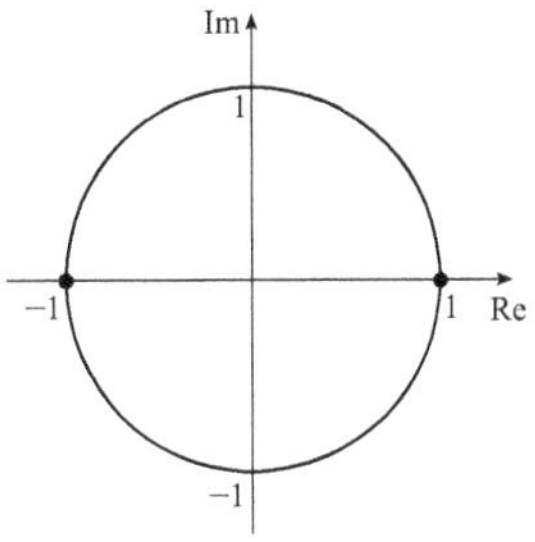

Ask about the fourth roots of 1, obtaining ± 1 and $\pm i$, and draw them as well:

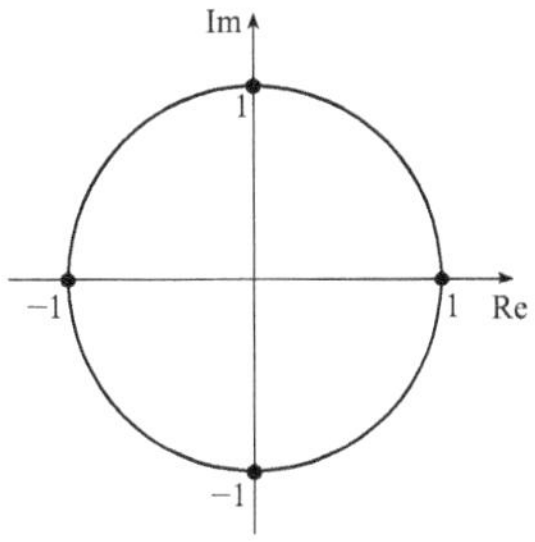

Now ask about the third roots of 1 and fifth roots of 1. you will probably have to do these as an example on the board. Draw them, and point out that the nth roots of one are always equally spaced around the unit circle.

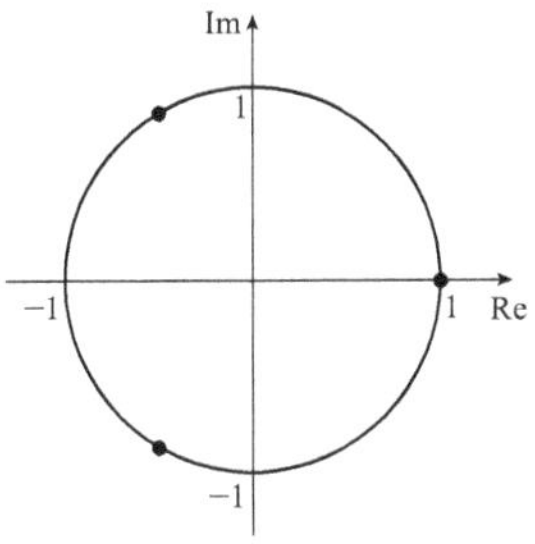

Third roots of 1

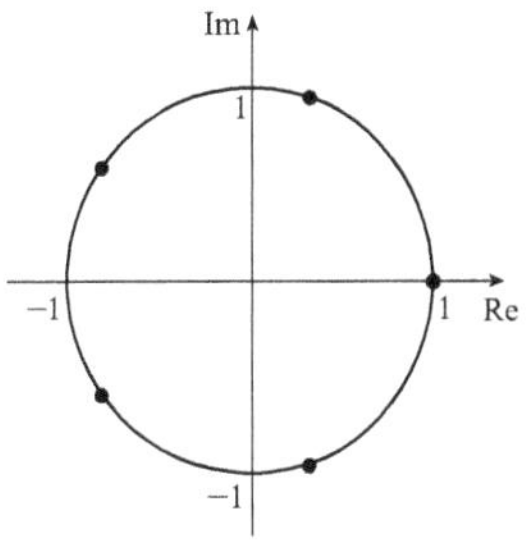

Fifth roots of 1

▼ Examples

- Switching between forms:
  - $4\left(\cos \frac{5\pi}{4} + i \sin \frac{5\pi}{4}\right) = -2\sqrt{2} - 2\sqrt{2}i$
  - $4\left(\cos 1 + i \sin 1\right) \approx 2.161 + 3.366i$
  - $3\left(\cos 20^\circ + i \sin 20^\circ\right) \approx 2.81907786 + 1.02606043i$
- $(4 - 3i)^5 = -3116 + 237i$

▼ Group Work: The Complex Pirates

This whimsical activity allows the students to practice adding and subtracting complex numbers in various forms.

Answer:

After step 3: $100(\cos 90° + i\sin 90°) = 100i$

After step 5: $100i + (300\cos 130° + i\sin 130°) \approx -192.8 + 329.8i$

After step 7: $-192.8 + 329.8i + 300(\cos 50° + i\sin 50°) \approx 559.6i$

After step 9: $559.6i + 700(\cos 0° + i\sin 0°) \approx 700 + 559.6i$

After step 11: $700 + 559.6i + 100(\cos 20° + i\sin 20°) \approx 794.0 + 594.0i$

Distance from the old skull: $\sqrt{794^2 + 594^2} \approx 991.5$ ft

Direction: $\arctan \frac{594}{794} \approx 36.8°$, just east of northeast

▼ Homework Problems

Core Exercises: 4, 9, 16, 22, 25, 32, 54, 61, 70, 87

Sample Assignment: 4, 9, 12, 16, 18, 19, 22, 25, 32, 43, 50, 54, 58, 61, 67, 70, 73, 87, 94

GROUP WORK, SECTION 5.3

The Complex Pirates

The fearsome pirate Rob Rackham has given you his treasure map for safekeeping. Ha ha ha! Silly Rob Rackham!

You get to Bulbous Island and find your way to the old skull, and then unroll the map to take it from there:

1. Get ye to Bulbous Island.
2. Get ye to ye Olde Skull.
3. Go North 100 feet.
4. Turn 40 degrees counterclockwise.
5. Walk 300 feet forward.
6. Turn 80 degrees clockwise.
7. Walk 300 feet forward.
8. Say "Arrrr!" rather loudly.
9. Turn to face East. Walk 700 feet due east.
10. Turn 20 degrees counterclockwise
11. Walk 100 feet forward.
12. Hop on one foot.
13. Dig.

How far away is the treasure from the old skull, and in what direction does it lie?

5.4 PLANE CURVES AND PARAMETRIC EQUATIONS

▼ Suggested Time and Emphasis

1 class. Recommended material.

▼ Points to Stress

1. Definition of parametric equations.
2. Sketching parametric curves.
3. Eliminating the parameter.
4. Polar equations in parametric form.

▼ Sample Questions

- **Text Question:** What is the difference between a function and a parametric curve?

  Answer: Many answers are possible. The graph of a function can be made into a parametric curve, but not necessarily the other way around. A function has to pass the vertical line test and a parametric curve does not.

- **Drill Question:** Sketch the parametric curve $x(t) = \sin t \; y(t) = t^2$, $0 \le t \le \pi$. Is the point $(1, \frac{\pi}{4})$ on this curve?

  Answer:

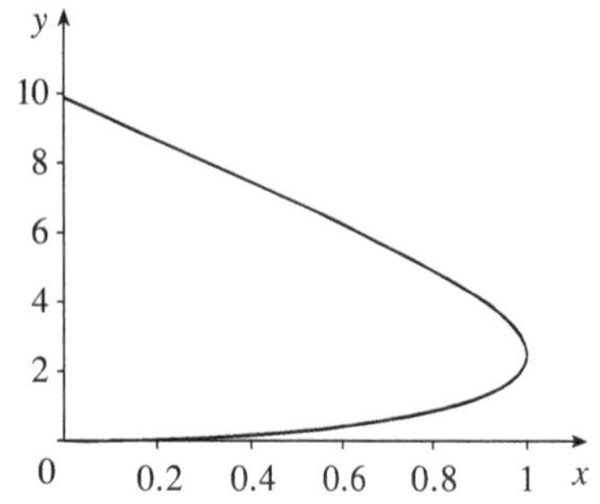

  $(1, \frac{\pi}{4})$ is not on the curve.

▼ In-Class Materials

- Discuss the process of going from a parametric curve to a relation between x and y. The figure at right is meant to help students see how parametrized curves are sketched out over time. First sketch $(x(t), y(t))$, starting at the initial point (the origin), and moving up and to the right. (Try to keep your speed constant.) Stop when the cycle is about to repeat. Then, to the right of the figure, graph the motion in the y-direction only. Then, below the figure, graph the motion in the x-direction. That graph is sideways because the x-axis is horizontal.

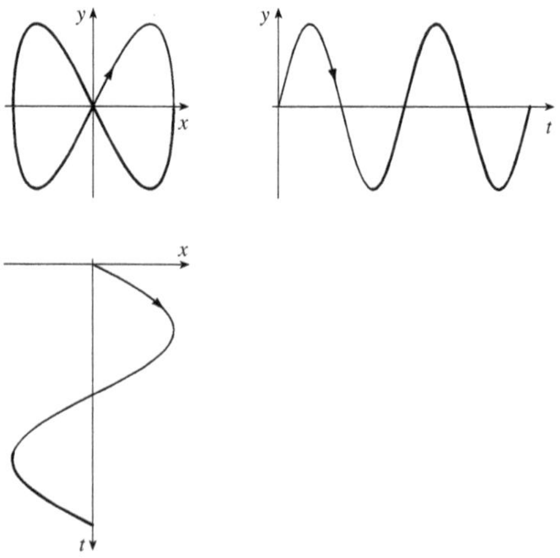

- Show how reversing the functions $x(t)$ and $y(t)$ yields the inverse of a given relation. For example, $\begin{cases} x(t) = t \\ y(t) = \sin t \end{cases}$ is the sine function, so $\begin{cases} x(t) = \sin t \\ y(t) = t \end{cases}$ is the general arcsine function. Display an inverse for $f(x) = x^3 + x + 2$ graphically using parametric equations. Explain the difficulties with the algebraic approach.
- Have the students get into groups, and have each group try to come up with the most interesting looking parametric curve. After displaying their best ones, perhaps show them the following examples:

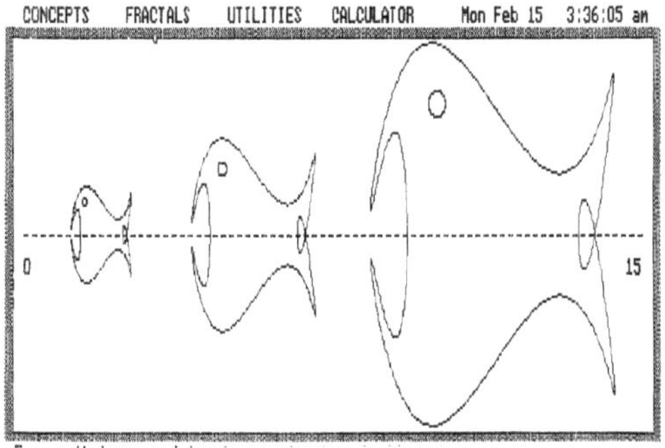

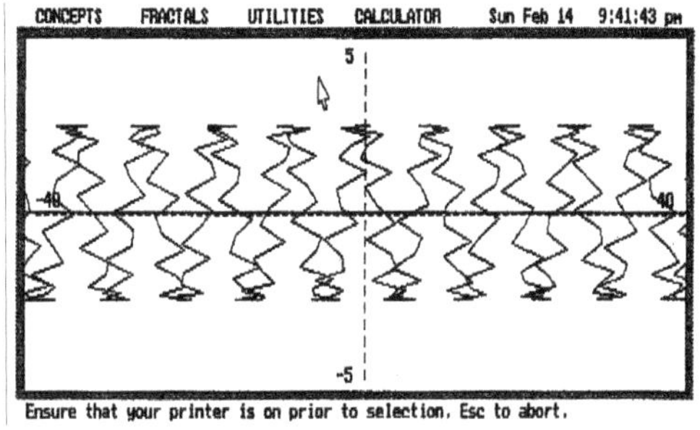

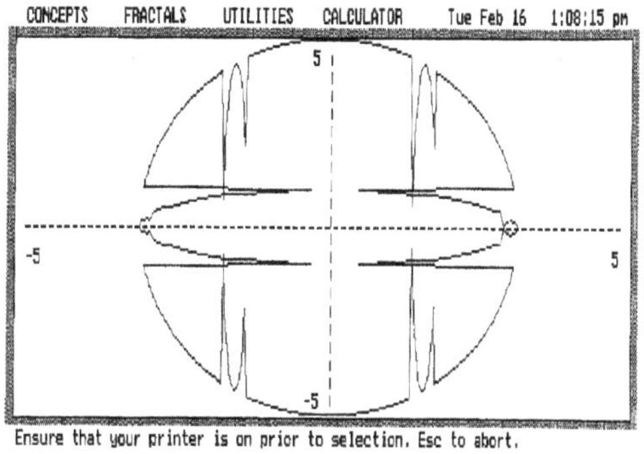

- Give an example of a curve such as $x(t) = \cos et$, $y(t) = \sin \sqrt{3}t$. This curve essentially fills the square $-1 \le x \le 1$, $-1 \le y \le 1$ in that the curve gets arbitrarily close to any point in the square. (It is not what mathematicians call a "space filling curve", because it does not actually hit every point in the square.) It can be simulated using a graphing calculator with the approximations $e \approx 2.7183$ and $\sqrt{3} \approx 1.7321$. The range $0 \le t \le 200$ should be sufficient to convey this property to the students. Next, describe the family of functions $x(t) = a \cos et$, $y(t) = b \sin \sqrt{3}t$. If the students are following well, perhaps consider the family $x(t) = \cos at$, $y(t) = \sin \sqrt{3}t$. The students might be tempted to conclude that this is always space-filling, but it is not for some values of a, such as $a = 2\sqrt{3}$.

▼ Examples

- Construction of an ellipse with center $(-1, -2)$ and axis lengths 2 and 3: We start with the circle $x = \cos\theta$, $y = \sin\theta$. We stretch it into shape: $x = 2\cos\theta$, $y = 3\sin\theta$. Then we move it into position: $x = 2\cos\theta - 1$, $y = 3\sin\theta - 2$.

▼ Group Work 1: Name that Parametrization

This exercise gives the students some practice playing with parametric curves. Before starting the activity, make sure that the students know how to graph a simple set of parametric curves using their technology. Make sure the students chose an appropriate value for `tstep`. If their graphs look jagged, there is a chance that their `tstep` is too large.

Answers:

1. (a) Every time $\sin t$ completes one cycle, $\cos 2t$ completes two cycles.
   (b) $y = (1 - x^2) - x^2 = 1 - 2x^2$
3. They are the same if considered as curves in the plane. The second one "moves" twice as fast as the first.
4. (a) The curves are inverses of each other (or, they are symmetric about the line $y = x$).
   (b) The curves are inverses of each other (or, they are symmetric about the line $y = x$).
5. (a) We graph f and see that it passes the Horizontal Line Test.
   (b)

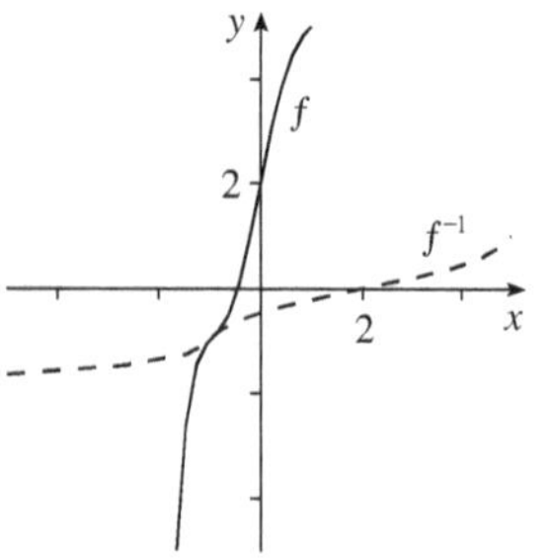

▼ Group Work 2: How Many Ways Can You Trace That Curve?

This exercise gives students practice finding different parametrizations for the circle $x^2 + y^2 = 4$, using the forms $x(t) = a_1 \cos a_2 t$, $y(t) = a_3 \sin a_4 t$. Students may use a graphing calculator.

Answers:

1. Each of a_1 and a_3 must be ± 2.
2. Once around counterclockwise, starting at $(1, 0)$.
3. $a_2 = a_4 = 2$, $a_2 = a_4 = 3$
4. Once around clockwise, starting at $(1, 0)$.
5. $a_2 = 3$, $a_4 = -3$; $a_2 = 5$, $a_4 = -5$

▼ Homework Problems

Core Exercises: 3, 14, 27, 30, 34, 37, 54, 55, 66

Sample Assignment: 3, 8, 14, 23, 27, 30, 32, 34, 37, 39, 50, 53, 54, 55, 56, 66

GROUP WORK 1, SECTION 5.4

Name that Parametrization

1. Consider the graph of the following set of parametric equations:

$$x(t) = \sin t \qquad y(t) = \cos 2t \qquad 0 < t < \infty$$

(a) Graph this curve with your calculator. Why does it look the way it does?

(b) Write this equation in the form $y = f(x)$. (*Hint:* Use the formula $\cos 2\theta = \cos^2\theta - \sin^2\theta$.)

2. Try to guess what the graph of the following set of parametric equations looks like, and then see if you are right.

$$x(t) = \sin 2t \qquad y(t) = \cos 6t \qquad 0 \le t \le 4\pi$$

These curves are called Lissajous figures, and are used in electrical engineering to see if two signals are "in sync".

3. Consider these two sets of parametric equations:

$$x(t) = t \qquad y(t) = \sin t \qquad 0 < t < \infty$$

$$x(t) = 2t \qquad y(t) = \sin 2t \qquad 0 < t < \infty$$

What is the relationship between their associated curves?

4. (a) Consider these two sets of parametric equations:

$$x(t) = t \qquad y(t) = \sin t \qquad 0 < t < \infty$$

$$x(t) = \sin t \qquad y(t) = t \qquad 0 < t < \infty$$

What is the difference between their associated curves?

(b) Given any set of equations of the form

$$x(t) = t \qquad y(t) = f(t) \qquad 0 < t < \infty$$

What does the graph of the set of equations

$$y(t) = t \qquad x(t) = f(t) \qquad 0 < t < \infty$$

look like?

5. (a) Use a graphing calculator to check that $f(x) = x^5 - 3x^3 + 5x + 2$ is one-to-one.

(b) Graph its inverse function f^{-1}.

GROUP WORK 2, SECTION 5.4

How Many Ways Can You Trace That Curve?

Consider the circle $x^2+y^2=4$. We want to construct parametric curves $x(t)=a_1\cos a_2 t$, $y(t)=a_3\sin a_4 t$, with a_1, $a_2>0$, which will trace this circle in different ways.

1. What must be the values of a_1 and a_3 so that $(x(t), y(t))$ lies on the circle $x^2+y^2=4$?

2. Describe the motion of the particle if you set $a_2=a_4=1$ and let $0\le t\le 2\pi$. What is the starting point?

3. What choice of a_2 and a_4 will trace the circle twice in a counterclockwise direction starting at $(1,0)$? What choice will trace the circle three times in a counterclockwise direction?

4. Describe the motion if you set $a_2=1$, $a_4=-1$ and let $0\le t\le 2\pi$.

5. What choice of a_2 and a_4 will trace the circle three times in a clockwise direction starting at $(1,0)$? What choice will trace the circle five times in a clockwise direction?

6 VECTORS IN TWO AND THREE DIMENSIONS

6.1 VECTORS IN TWO DIMENSIONS

▼ Suggested Time and Emphasis

1 class. Essential material.

▼ Points to Stress

1. The difference between a vector and a scalar.
2. Translational invariance of vectors, including the concept of standard position.
3. The geometric interpretation of vector addition and scalar multiplication.
4. The magnitude of a vector, its direction, and the resolution of a vector into its components.

▼ Sample Questions

- **Text Question:** What are the components of the vector from $(3,3)$ to $(4,4)$?

  Answer: 1 and 1
- **Drill Question:** What are the magnitude and direction of the vector $3\mathbf{i}+3\mathbf{j}$?

  Answer: Magnitude $3\sqrt{2}$, direction $\frac{\pi}{4}$

▼ In-Class Materials

- Assume $\mathbf{v}_1$ and $\mathbf{v}_2$ are in standard position. Demonstrate that the vector $\mathbf{v}_3=\mathbf{v}_1-\mathbf{v}_2$ can be viewed as the vector with initial point $\mathbf{v}_2$ and terminal point $\mathbf{v}_1$. Draw this picture in $\mathbb{R}^2$ and (if you can) in $\mathbb{R}^3$.
- Ask students to find r and s such that $\langle 3,2\rangle = r\langle 1,1\rangle + s\langle 0,4\rangle$. Repeat for $\langle 3,2\rangle = r\langle 1,1\rangle - s\langle 0,4\rangle$. Ask students if there is a vector $\langle x,y\rangle$ such that we cannot find r and s with $\langle x,y\rangle = r\langle 1,1\rangle + s\langle 0,4\rangle$. Note that this kind of question becomes very important in linear algebra.
- Go over the properties of vectors from both algebraic and geometric perspectives.

▼ Examples

- Geometric and algebraic addition and subtraction:

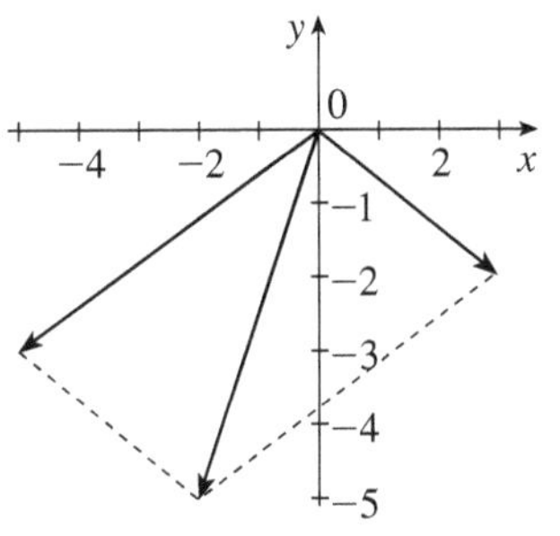

$\langle 3,-2\rangle + \langle -5,-3\rangle = \langle -2,-5\rangle$

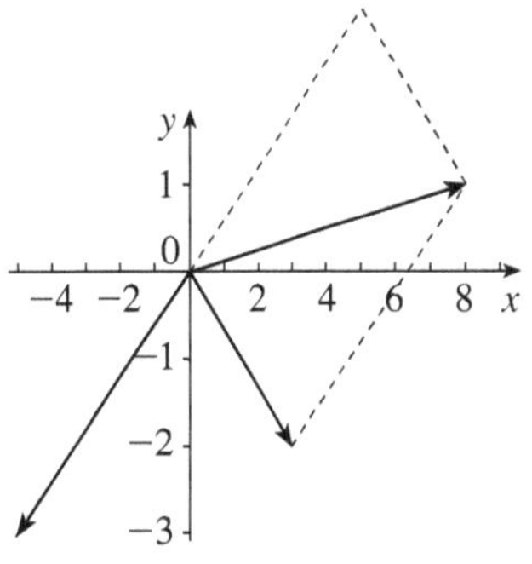

$\langle 3,-2\rangle - \langle -5,-3\rangle = \langle 8,1\rangle$

▼ Group Work 1: Vector or Scalar?

This activity can be done with the whole class simultaneously. Allow some time to work individually, and then have them vote on one question at a time.

Answers:

Quantity	Vector?	Scalar?
Speed		×
Force	×	
Debt		×
Velocity	×	
Acceleration	×	×
Momentum	×	

Quantity	Vector?	Scalar?
Energy		×
Temperature		×
Work	×	
Friction	×	
Current		×

▼ Group Work 2: Where Do They Point?

This group work extends the idea of adding several vectors without coordinates. For Problem 1, consider giving the hint to first try computing $\mathbf{a} + \mathbf{b} + \mathbf{c} + \mathbf{d} + \mathbf{e}$.

Answers:

1. $-2\mathrm{e}$

2. (a) $\left\langle -\frac{1}{2}r, -\frac{\sqrt{3}}{2}r \right\rangle$ **(b)** Since $\mathbf{a} + \mathbf{b} = \langle r, 0 \rangle$, we have $\mathbf{a} + \mathbf{b} + \mathbf{c} + \mathbf{d} = -(\mathbf{a} + \mathbf{d}) = \langle -r, 0 \rangle$.

▼ Group Work 3: Cones in General

This activity will introduce students to the concept of the cone of a set of vectors. You might want to stop after the first question and make sure everyone understands the definition, or give a different example to do all together before setting them loose.

Answers

1.

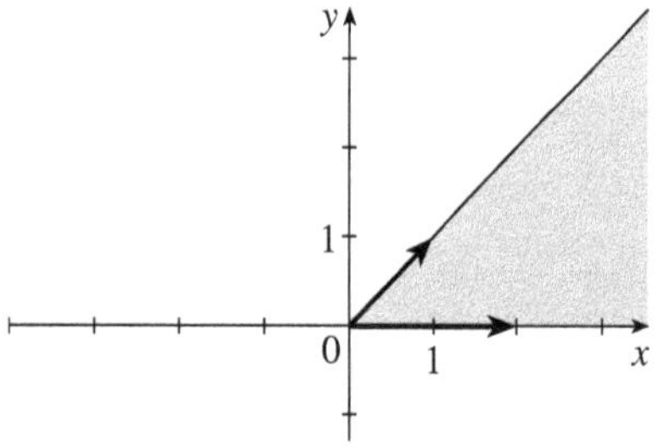

2.

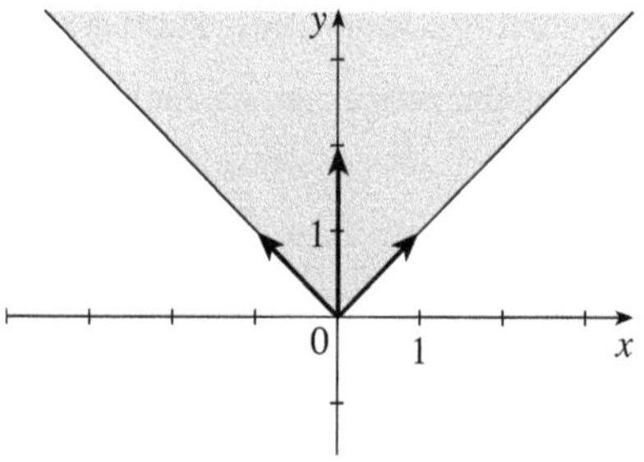

3. Yes. $\langle 0, 7, 3 \rangle = 2 \langle 1, 1, 1 \rangle + 2 \langle -1, 2, 1 \rangle + (-1) \langle 0, -1, 1 \rangle$

4. Yes. $\langle 0, 7, 3 \rangle = 2 \langle 1, 1, 1 \rangle + 2 \langle -1, 2, 1 \rangle + (-1) \langle 0, -1, 1 \rangle + 0 \langle 0, 1, 0 \rangle$

▼ Homework Problems

Core Exercises: 5, 12, 20, 28, 31, 38, 39, 42, 51, 53

Sample Assignment: 5, 6, 12, 16, 20, 23, 28, 31, 34, 38, 39, 42, 45, 48, 51, 53, 60

GROUP WORK 1, SECTION 6.1

Vector or Scalar?

Some of the following quantities are vectors, and some are scalars. Classify them all by checking the appropriate box, and then come up with two or three with which to stump your classmates, and hopefully your instructor.

Quantity	Vector?	Scalar?
Speed		
Force		
Debt		
Velocity		
Acceleration		
Momentum		
Energy		
Temperature		
Work		
Friction		
Current		

GROUP WORK 2, SECTION 6.1

Where Do They Point?

1. Compute $\mathbf{a}+\mathbf{b}+\mathbf{c}+\mathbf{d}-\mathbf{e}$ for the following diagram.

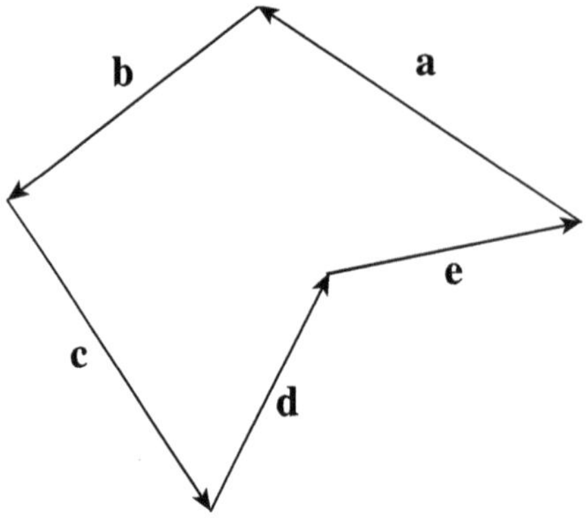

2. (a) Compute the position representation $\overrightarrow{OP}$ for $\mathbf{a}+\mathbf{b}$ in the following diagram. Give the coordinates of the point P.

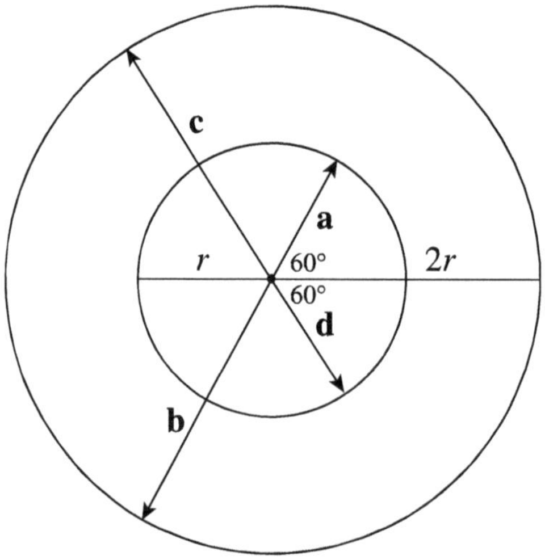

(b) Compute $\mathbf{a}+\mathbf{b}+\mathbf{c}+\mathbf{d}$.

GROUP WORK 3, SECTION 6.1

Cones in General

The **cone** in $\mathbb{R}^n$ generated by a collection of vectors $\mathbf{v}_1, \mathbf{v}_2, \mathbf{v}_3, \ldots, \mathbf{v}_k$ is the set of all nonnegative linear combinations of these vectors, drawn in standard position. (A nonnegative linear combination is of the form $\sum_{i=1}^{k} c_i \mathbf{v}_i$, where $c_i \geq 0$, that is, a linear combination where all of the coefficients are nonnegative.)

1. Cones in $\mathbb{R}^2$ are easy to identify. Shade in the cone generated by $\mathbf{v}_1 = \langle 1, 1 \rangle$ and $\mathbf{v}_2 = \langle 2, 0 \rangle$.

2. Shade in the cone generated by $\mathbf{v}_1 = \langle -1, 1 \rangle$, $\mathbf{v}_2 = \langle 0, 2 \rangle$, and $\mathbf{v}_3 = \langle 1, 1 \rangle$.

3. Does the vector $\langle 0, 7, 3 \rangle$ belong to the cone generated by $\langle 1, 1, 1 \rangle$, $\langle -1, 2, 1 \rangle$, and $\langle 0, -1, 1 \rangle$? Why or why not?

4. Does the vector $\langle 0, 7, 3 \rangle$ belong to the cone generated by $\langle 1, 1, 1 \rangle$, $\langle -1, 2, 1 \rangle$, $\langle 0, -1, 1 \rangle$, and $\langle 0, 1, 0 \rangle$? Why or why not?

6.2 THE DOT PRODUCT

▼ Suggested Time and Emphasis

$\frac{1}{2}$–1 class. Essential material.

▼ Points to Stress

1. The dot product and its properties.
2. Definition of orthogonality.
3. Length of a vector.
4. Projections and their applications to computer science.

▼ Sample Questions

- **Text Question:** Compute $\langle 3, 4 \rangle \cdot \langle -1, 2 \rangle$.

  Answer: 5

- **Drill Question:** Consider the following pairs of vectors, all of which have length 1:

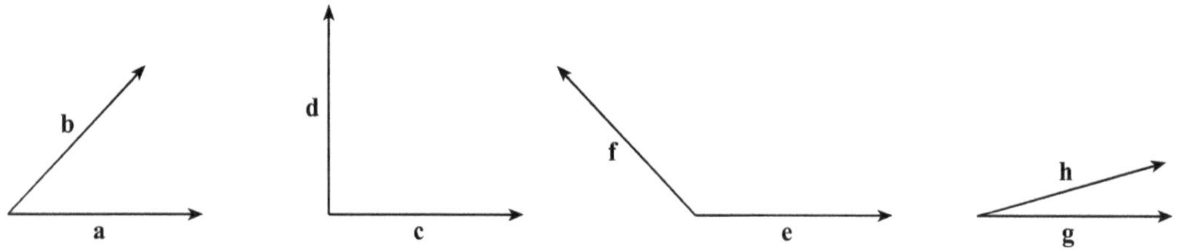

  Put the following quantities in order, from smallest to largest:

  $$\mathbf{a} \cdot \mathbf{b} \qquad \mathbf{c} \cdot \mathbf{d} \qquad \mathbf{e} \cdot \mathbf{f} \qquad \mathbf{g} \cdot \mathbf{h}$$

  Answer: $\mathbf{e} \cdot \mathbf{f}, \mathbf{c} \cdot \mathbf{d}, \mathbf{a} \cdot \mathbf{b}, \mathbf{g} \cdot \mathbf{h}$

▼ In-Class Materials

- Assume that $\mathbf{a} \cdot \mathbf{b} = \mathbf{a} \cdot \mathbf{c}$ and $\mathbf{a} \neq 0$. Pose the question, "Is it necessarily true that $\mathbf{b} = \mathbf{c}$?" When you've convinced them (perhaps by example) that the answer is "no", the next logical question to ask is, "What can we say about $\mathbf{b}$ and $\mathbf{c}$?" It can be shown that $\mathbf{b}$ and $\mathbf{c}$ have the same projection onto $\mathbf{a}$, since $\mathbf{a} \perp (\mathbf{b} - \mathbf{c})$.

- Demonstrate the distributive property in general. Note that while the dot product is commutative and distributive, the associative property makes no sense, as it is not possible to take the dot product of three vectors.

- Demonstrate the proper formation of statements involving dot products. For example, the statement $c(\mathbf{a} \cdot \mathbf{b})$ makes sense, while the statements $\mathbf{d} \cdot (\mathbf{a} \cdot \mathbf{b})$ and $c \cdot \mathbf{a}$ do not.

- This is a nice, direct application of vector projections: It is clear that a weight will slide more quickly down ramp 2 than down ramp 1:

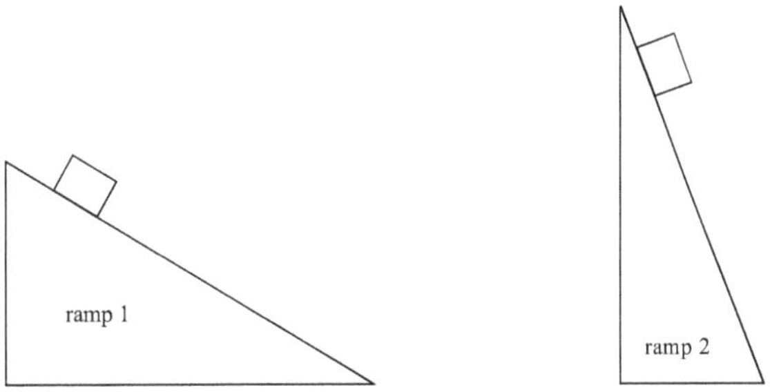

Gravity is the same in both cases, yet there is a definite difference in speed. The reason behind this is interesting. Gravity is doing two things at once: it is letting the weight slide down, and it is also preventing the weight from floating off the ramp and flying into outer space. We can draw a "free body diagram" that shows how the gravity available to let the weight slide down is affected by the angle of the ramp.

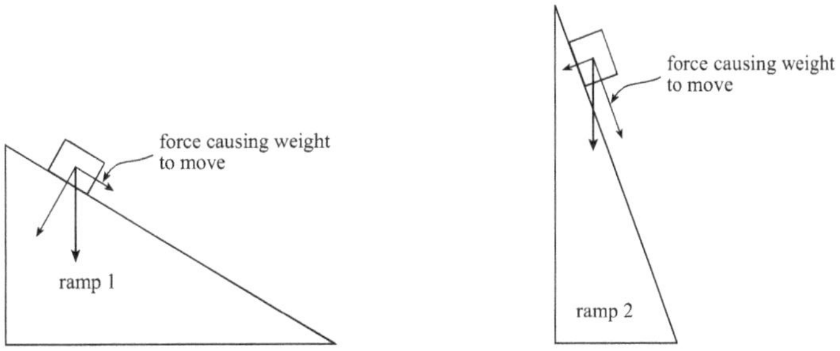

A block slides faster on a steeper slope because the projection of the gravitational force in the direction of the slope is larger. There is more force pushing the block down the slope, and less of a force holding it to the surface of the slope.

▼ Examples

- $\langle 5, 6, 2\rangle \cdot \langle -3, 1, 0\rangle = -9$
- Two vectors that are orthogonal: $\mathbf{a} = \langle 5, 6, 2\rangle$, $\mathbf{b} = \left\langle 1, -1, \frac{1}{2}\right\rangle$
- Projections: If $\mathbf{a} = \langle 2, 1, -1\rangle$ and $\mathbf{b} = \langle 3, 2, 7\rangle$, then $\operatorname{proj}_{\mathbf{a}} \mathbf{b} = \left\langle \frac{1}{3}, \frac{1}{6}, -\frac{1}{6}\right\rangle$, $\operatorname{proj}_{\mathbf{b}} \mathbf{a} = \left\langle \frac{3}{62}, \frac{1}{31}, \frac{7}{62}\right\rangle$, and $|\mathbf{a}| = \sqrt{6}$.

▼ Group Work 1: Orthogonal Projections Rule

This worksheet extends what the students know about orthogonal projections to $\mathbb{R}^3$. If they have a strong conceptual knowledge of projections, this should come easily.

Answers:

1. P is a plane through the origin.
2. Answers will vary.
3. $\langle u_1 - c_1 - c_2, u_2 - c_1, u_3 - c_1\rangle$
4. $-3c_1 - c_2 + u_1 + u_2 + u_3 = 0$, $u_1 - c_1 - c_2 = 0$
5. $c_1 = \frac{1}{2}(u_1 + u_2)$, $c_2 = \frac{1}{2}(2u_1 - u_2 - u_3)$

▼ Group Work 2: The Right Stuff

Give each group of students a different set of three points, and have them use vectors to determine if they form a right triangle. (It is easiest to write the points directly on the pages before handing them out.) They can do this using dot products, by calculating side lengths and using the Pythagorean Theorem, or by calculating the slopes of lines between the pairs of points. Have the students whose points are in $\mathbb{R}^2$ carefully graph their points to provide a visual check. At the end of the exercise, point out that using the dot product is the easier method.

Sample Answers:

$(-2,-1), (-2,8), (8,-1)$	Right	$(3,4), (3,12), (6,5)$	Not right
$(0,0), (10,7), (-14,20)$	Right	$(2,1,2), (3,3,1), (2,2,4)$	Right
$(-1,-2,-3), (0,0,-4), (-1,-1,-1)$	Right	$(2,3,6), (3,4,7), (3,3,6)$	Right

▼ Group Work 3: The Regular Hexagon

If the students have trouble with this one, copy the figure onto the blackboard. Then draw a point at its center, and draw lines from this point to every vertex. This modified figure should make the exercise more straightforward.

Answers:

1. 1, 1, 1

2. $120°$

3. $\cos 60° = \frac{1}{2}$

4. $-\frac{1}{2}$

5. $\left\langle -\frac{1}{2}, 0 \right\rangle, \left\langle -\frac{1}{4}, \frac{\sqrt{3}}{4} \right\rangle$

6. 1

▼ Homework Problems

Core Exercises: 1, 6, 9, 17, 22, 26, 31, 35, 42, 46

Sample Assignment: 1, 6, 9, 12, 14, 17, 19, 22, 26, 28, 31, 33, 35, 38, 42, 44, 46, 49

GROUP WORK 1, SECTION 6.2

Orthogonal Projections Rule

We know how to orthogonally project one vector onto another. Let's try to extend this procedure in the setting of $\mathbb{R}^3$.

1. Let $\mathbf{v}_1 = \langle 1, 1, 1 \rangle$ and $\mathbf{v}_2 = \langle 1, 0, 0 \rangle$ be vectors (in standard position) in $\mathbb{R}^3$ and let P denote the set of all linear combinations of $\mathbf{v}_1$ and $\mathbf{v}_2$. Describe what P looks like.

2. With P fixed, we want to orthogonally project a given vector $\mathbf{u}$ onto P. That is, we want the orthogonal component of $\mathbf{u}$ that belongs to P. Sketch P and orthogonal projections for a couple of different $\mathbf{u}$'s. Let's let $\operatorname{proj}_P \mathbf{u}$ denote the projection.

3. Now since $\operatorname{proj}_P \mathbf{u}$ belongs to P, we can write $\operatorname{proj}_P \mathbf{u} = c_1\mathbf{v}_1 + c_2\mathbf{v}_2$ for some constants c_1 and c_2. Our goal is to obtain formulas for c_1 and c_2. Expressing $\mathbf{u}$ using coordinates, we write $\mathbf{u} = \langle u_1, u_2, u_3 \rangle$. Now express, in coordinate form, the vector $\mathbf{u} - \operatorname{proj}_P \mathbf{u}$. Make sure to use the equation for $\operatorname{proj}_P \mathbf{u}$ and the coordinates of $\mathbf{v}_1$ and $\mathbf{v}_2$ above.

4. From our work in Problem 2, we know that $\mathbf{u} - \operatorname{proj}_P \mathbf{u}$ is orthogonal to both $\mathbf{v}_1$ and $\mathbf{v}_2$. Combine this fact with your work from Problem 3. to find 2 equations involving the unknowns c_1 and c_2.

5. Finally, solve the above equations to give formulas for c_1 and c_2.

GROUP WORK 2, SECTION 6.2

The Right Stuff

Consider the points (,), (,), and (,). These three points form a triangle. Is this triangle a right triangle? Justify your answer.

GROUP WORK 3, SECTION 6.2

The Regular Hexagon

Consider the following regular hexagon:

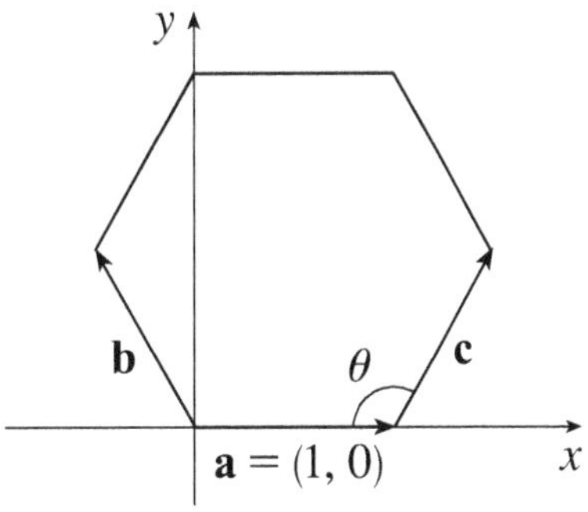

1. Compute $|\mathbf{a}|$, $|\mathbf{b}|$, and $|\mathbf{c}|$.

2. What is the angle θ?

3. What is $\mathbf{a} \cdot \mathbf{c}$?

4. What is $\mathbf{a} \cdot \mathbf{b}$?

5. Compute $\text{proj}_{\mathbf{a}}\, \mathbf{b}$ and $\text{proj}_{\mathbf{b}}\, \mathbf{c}$.

6. Compute the x-component of $\mathbf{a} + \mathbf{b} + \mathbf{c}$.

6.3 THREE-DIMENSIONAL COORDINATE GEOMETRY

▼ Suggested Time and Emphasis

$\frac{1}{2}$–1 class. Essential material.

▼ Points to Stress

1. Points in three-dimensional space, including notation and the distance formula.

2. Equations of planes parallel to one of the coordinate planes.

3. The equation of a sphere.

▼ Sample Questions

- **Text Question:** Explain why $x = 3$ is the equation of a plane, not a line, in three-dimensional space.

  Answer: Because the coordinates y and z are not specified, they can take on any value, extending the two-dimensional line $x = 3$ into a three-dimensional plane.
- **Drill Question:** What is the distance between the points $(-1, -1, -1)$ and $(-4, -1, 3)$?

  Answer: $\sqrt{[-1-(-4)]^2 + [-1-(-1)]^2 + (-1-3)^2} = 5$

▼ In-Class Materials

- If possible, mark off one corner of the lecture room with electrical tape. Determine the coordinates of various objects in the room and/or find the equation of the plane of the chalkboard.
- Explain why the equation $z = \sqrt{R^2 - (x^2 + y^2)}$ represents the hemisphere $z \geq 0$, whereas the equation $x^2 + y^2 + z^2 = R^2$ represents a full sphere, perhaps making a two-dimensional analogy.
- Introduce the equation of a circular cylinder as an extension to $\mathbb{R}^3$ of the equation of the circle $x^2+y^2 = r^2$. Then describe some cylindrical surfaces such as $y = x^2$, $x^2 - y^2 = 1$ and $z = \sin y$.
- Describe the set of points whose distance from each of $(0, 0, 1)$ and $(0, 0, -1)$ is 1. Then similarly discuss the set of points whose distance from these points is 2, and then 0.5. Have the students explain why the set of all points equidistant in space from two given points is a plane.

▼ Examples

- One sphere in three forms:

  1. The sphere with center $(1, 2, 3)$ and radius 4.
  2. The sphere with equation $(x-1)^2 + (y-2)^2 + (z-3)^2 = 16$.
  3. The sphere with equation $x^2 - 2x + y^2 - 4y + z^2 - 6z - 2 = 0$.

▼ Group Work 1: Working with Surfaces in Three-Dimensional Space

Visualizing and sketching surfaces in three-dimensional space can be difficult for students. It is recommended that the answer key (included following the question sheet) be distributed after the students have had time to work.

▼ Group Work 2: Lines, Lines, Everywhere Lines

The idea is to help students visualize the triangular surface $x + y + z = 1$, $x, y, z \geq 0$. Note that at this stage, the students have not yet learned about the general equations of lines and planes. Part 3 is ambitious, and requires some formal thinking. Depending on the mathematical maturity of your class, you may want to either give a hint to the students, or perhaps do that part as a class. Parts 3 and 4 are printed on a separate page, so you can decide what to do based upon how the students are doing on the first two parts.

Answers:

1. In the xy-plane, we have $z = 0$. If $x + y = 1$ then $x + y + 0 = 1$.

2. As in Problem 1, *mutatis mutandis*.

3. Adding the equations $x_1 + y_1 + z_1 = 1$ and $x_2 + y_2 + z_2 = 1$, we have $(x_1 + x_2) + (y_1 + y_2) + (z_1 + z_2) = 2$, so $\frac{(x_1 + x_2)}{2} + \frac{(y_1 + y_2)}{2} + \frac{(z_1 + z_2)}{2} = 1$, showing that the midpoint is in S.

4.

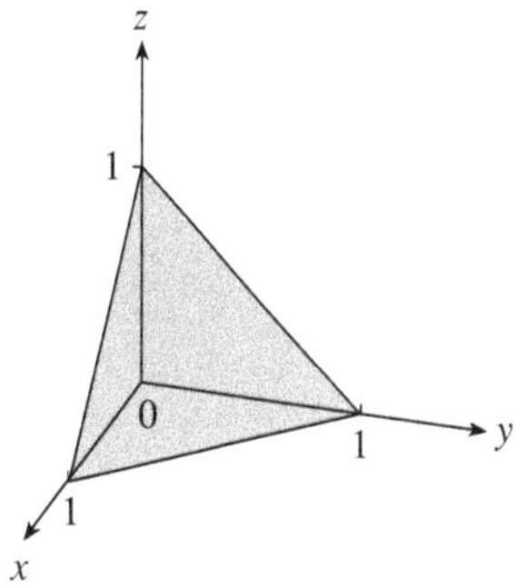

▼ Group Work 3: Fun with Visualization

Ask the students to turn off their calculators and put down their pencils. Start by asking them to picture a plane in $\mathbb{R}^3$. Ask them how many regions in space are formed by a plane. (Answer: Two regions)

Then ask them how many regions are formed by *two* planes. Notice that there are two possible answers here. If the planes are parallel, they divide space into three regions, like a layer cake. If they cross, then four regions are formed. Put the students into groups and hand out the activity. Give them time to argue with each other, and perhaps foster inter-group discussion as well. After the activity is over, discuss the analogy between dividing space by planes and dividing a plane by lines. This activity can be shortened by cutting out Problems 4 and 5.

Answers:

1. 2 **2.** 3 or 4 **3.** 8 **4.** 4, 6, 7, or 8 **5.** 15

▼ Homework Problems

Core Exercises: 1, 4, 8, 12, 15, 21, 23

Sample Assignment: 1, 4, 5, 8, 9, 12, 15, 18, 20, 21, 22, 23

GROUP WORK 1, SECTION 6.3

Working with Surfaces in Three-Dimensional Space

Draw (or describe in words) the surfaces described by the following five equations. The best way to do this exercise is to think first about the surfaces, before you calculate anything.

1. $x^2 + y^2 = 3^2$

2. $y^2 + z^2 = 1$

3. $z = y^2$

4. $xy = 1$

5. $x^2 + y^2 = z$

GROUP WORK 1, SECTION 6.3

Working with Surfaces in Three-Dimensional Space (Solutions)

1. $x^2 + y^2 = 3^2$

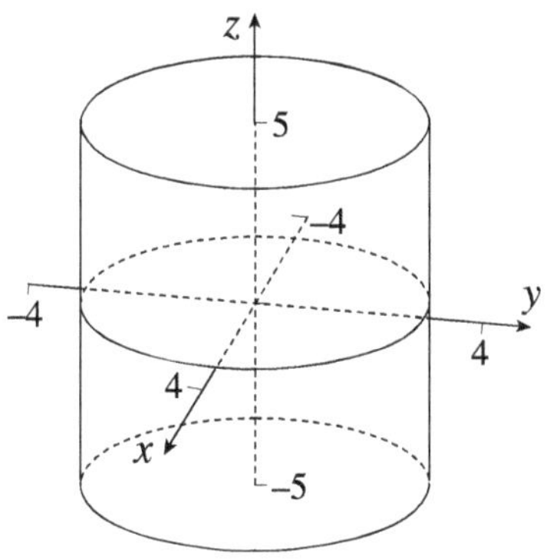

This is a right circular cylinder of radius 3 with axis the z-axis.

2. $y^2 + z^2 = 1$

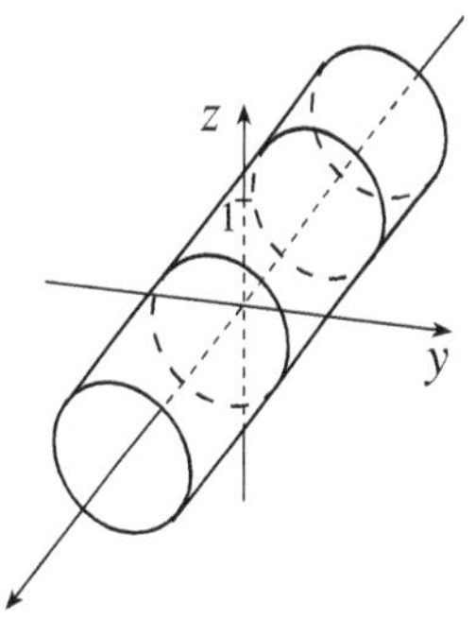

This is a right circular cylinder of radius 1 with axis the x-axis.

3. $z = y^2$

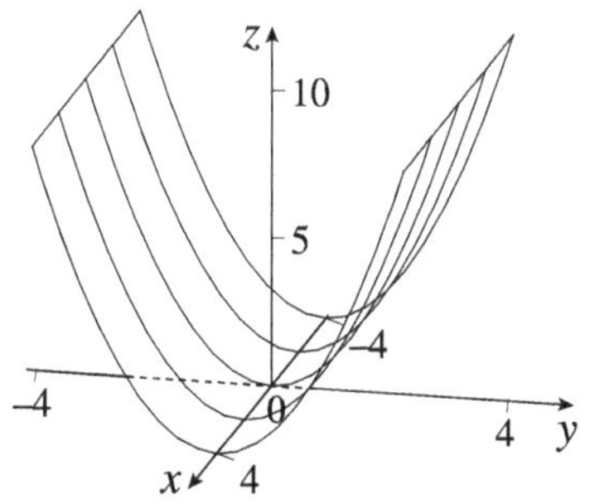

This is a parabolic cylinder parallel to the x-axis.

4. $xy = 1$

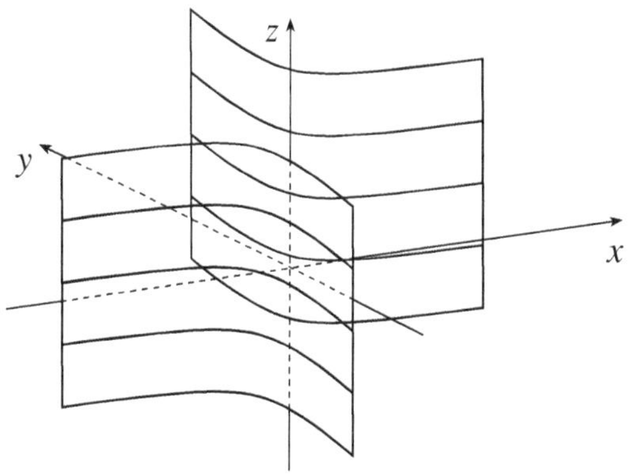

This is a hyperbolic cylinder parallel to the z-axis.

5. $x^2 + y^2 = z$

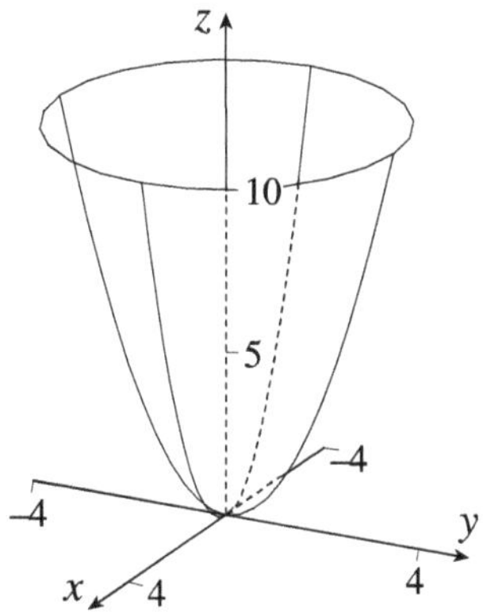

This is a paraboloid opening upward.

GROUP WORK 2, SECTION 6.3

Lines, Lines, Everywhere Lines

Consider the surface S: $x + y + z = 1$, $x \geq 0$, $y \geq 0$, $z \geq 0$.

1. Show that every point on the line segment in the xy-plane, $x + y = 1$ ($x \geq 0$, $y \geq 0$) is also a point on this surface.

2. Show that every point on the line segments $x + z = 1$ ($x \geq 0$, $z \geq 0$) and $y + z = 1$ ($y \geq 0$, $z \geq 0$) in the xz- and yz-planes are also points on this surface.

3. Given that two points $P_1\,(x_1, y_1, z_1)$ and $P_2\,(x_2, y_2, z_2)$ are on the surface S, show that their midpoint $P\left(\dfrac{x_1+x_2}{2}, \dfrac{y_1+y_2}{2}, \dfrac{z_1+z_2}{2}\right)$ must also be a point in S.

4. Draw a picture of S.

GROUP WORK 3, SECTION 6.3

Fun with Visualization

1. Picture a plane in $\mathbb{R}^3$. This plane divides space into how many regions?

2. How many regions can be formed by two planes?

3. What is the maximum number of regions that can be formed by three planes?

4. What are the other possible numbers of regions that can be formed by three planes?

5. What is the maximum number of regions that can be formed by four planes?

6.4 VECTORS IN THREE DIMENSIONS

▼ Suggested Time and Emphasis

$\frac{1}{2}$–1 class. Essential material.

▼ Points to Stress

1. The extension of vectors into three dimensions, including magnitude, vector addition, scalar multiplication, and the dot product.
2. The direction cosines of a vector in space.

▼ Sample Questions

- **Text Question:** If $\mathbf{a} = \langle a_1, a_2, a_3 \rangle$ has $a_2 > 0$ and $a_3 > 0$, then is the z-component of $-3\mathbf{a}$ positive or negative?
  Answer: Negative
- **Drill Question:** Compute the magnitude of $\langle 4, 13, 16 \rangle$.
  Answer: $\sqrt{4^2 + 13^2 + 16^2} = 21$

▼ In-Class Materials

- Picture two students standing on opposite sides of a giant cube, eight feet on a side, pushing towards the center. Have the students describe in what direction the cube will probably move. (Toward one of the two students, or not at all if the students have the same strength.) Now have them picture the students standing on adjacent sides, "kitty corner". Now have them describe the resultant motion, and try to get them to justify their answers. Lead into how vectors give us a natural way of describing the situation, and predicting the outcome.
- Show that it is tough to determine by inspection whether a three-dimensional triangle is a right triangle, but that three-dimensional vectors make it simple. For example, consider the triangle with vertices $(2, 1, 2)$, $(3, 3, 1)$, and $(2, 2, 4)$. Taking dot products of pairs of vectors between vertices, we find that one of these dot products is zero.
- Point out that although it is not easy to picture four-dimensional space, it is simple to generalize all our three-dimensional formulas and results. Write the four-dimensional vectors $\mathbf{u} = \langle 1, 2, -2, 6 \rangle$ and $\mathbf{v} = \langle 2, 1, 3, 0 \rangle$ on the board and have students figure out that the dot product is -2 without formally defining a four-dimensional dot product. Similarly, we can determine the magnitudes of four-dimensional vectors, their sums, differences, and scalar multiples.
- Find unit vectors in the directions of $\langle 8, 0, 0 \rangle$, $\langle 5, 5, 0 \rangle$ and $\langle 1, 2, 3 \rangle$, and explain what is happening geometrically. Then find a unit vector in the direction opposite that of $\langle -1, 1, 1 \rangle$.

▼ Examples

- Two perpendicular three-dimensional vectors: $\langle -2, 5, 1 \rangle$ and $\langle 5, 4, -10 \rangle$.

▼ Group Work: A Rectangular Solid

This activity foreshadows the process of resolving vectors into components and gives the students practice in working with vectors in three dimensions.

Answer: $\mathbf{a} = \langle -3, 4, 5 \rangle$, $\mathbf{b} = \langle 0, -4, 5 \rangle$, $\mathbf{c} = \langle -3, -4, 5 \rangle$

▼ Homework Problems

Core Exercises: 4, 7, 12, 13, 16, 26, 28, 29, 34, 42

Sample Assignment: 4, 5, 7, 9, 12, 13, 16, 17, 20, 24, 26, 28, 29, 34, 36, 39, 42, 43

GROUP WORK, SECTION 6.4

A Rectangular Solid

Consider the rectangular solid with one corner at $(0, 0, 0)$ and another at $(3, 4, 5)$, as shown below.

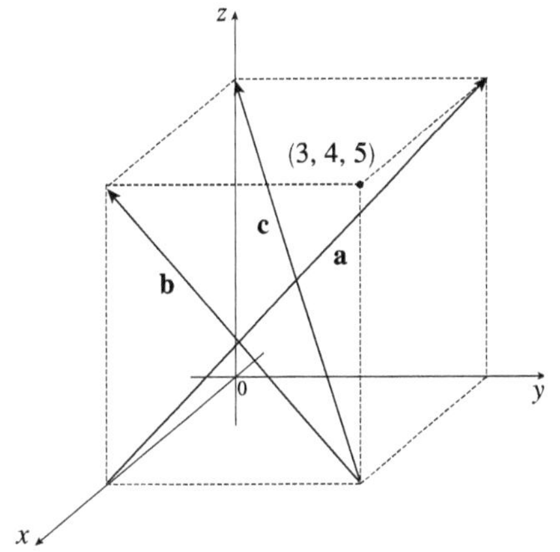

Find the component representations of each of the vectors labeled $\mathbf{a}$, $\mathbf{b}$, and $\mathbf{c}$.

6.5 THE CROSS PRODUCT

▼ Suggested Time and Emphasis

1 class. Essential material.

▼ Points to Stress

1. The cross product defined algebraically, and defined as a vector perpendicular to two given vectors, whose length is the area of the parallelogram determined by the vectors
2. The right-hand rule and properties of the cross product.

▼ Sample Questions

- **Text Question:** Why is it that if $\mathbf{a} \cdot (\mathbf{b} \times \mathbf{c}) = 0$ then $\mathbf{a}$, $\mathbf{b}$ and $\mathbf{c}$ are coplanar?

  Answer: $\mathbf{b} \times \mathbf{c}$ must be perpendicular to $\mathbf{a}$, and since $\mathbf{b} \times \mathbf{c}$ is also perpendicular to $\mathbf{b}$ and $\mathbf{c}$, we see that $\mathbf{a}$, $\mathbf{b}$, and $\mathbf{c}$ are coplanar.
- **Drill Question:** What is the volume of the parallelepiped defined by the vectors $\langle 1, 1, 1\rangle$, $\langle 0, 1, 1\rangle$, and $\langle 1, 2, 0\rangle$?

  Answer: 2

▼ In-Class Materials

- Pose the problem of finding a vector perpendicular to two given vectors (without considering the length of the resultant vector). Show that there is an obvious solution for $\langle 1, 0, 0\rangle$ and $\langle 0, 1, 0\rangle$. Then look at the two vectors $\langle 1, 0, 1\rangle$ and $\langle 0, 1, 1\rangle$. Repeat for $\langle -3, 1, -7\rangle$ and $\langle 0, -5, -5\rangle$. Point out that in this case an alternate algebraic solution using dot products gives $-3x + y - 7z = -5y - 5z$, $y = -z$, $x = -\frac{8}{3}z$, and so $\mathbf{c} = \left\langle -\frac{8}{3}, -1, 1\right\rangle$ works. Note that $\mathbf{c}$ is a scalar multiple of $\mathbf{a} \times \mathbf{b} = \langle -40, -15, -15\rangle$.
- Explain the geometry involved in computing the volume of a parallelepiped spanned by vectors $\mathbf{a}$, $\mathbf{b}$, and $\mathbf{c}$:

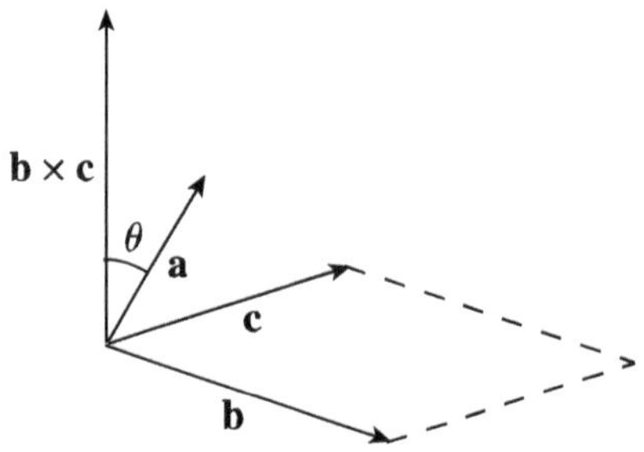

$$\begin{aligned} V &= (\text{area of the base}) \times (\text{height}) = |\mathbf{b} \times \mathbf{c}|\,\left|\text{proj}_{\mathbf{b}\times\mathbf{c}}\,\mathbf{a}\right| \\ &= |\mathbf{b} \times \mathbf{c}|\,|\mathbf{a}|\,|\cos\theta| = \mathbf{a} \cdot (\mathbf{b} \times \mathbf{c}) \end{aligned}$$

  Conclude that $\mathbf{a}$, $\mathbf{b}$, and $\mathbf{c}$ are coplanar if and only if the triple product $\mathbf{a} \cdot (\mathbf{b} \times \mathbf{c}) = 0$.
- Note that the correct pronunciation of the solid discussed above is parallelEPiped, a shortening of the original *parallelepipedon*, from the Greek words *para* (beside), *allele* (other), *epi* (upon), and *pedon* (ground). This reflects the fact that the solid always has a face parallel to the one on the ground.
- A common mistake made by right-handed students can be avoided if they are reminded to *put down their pencil* before using the right hand rule!

- Point out that while $\mathbf{a} \times \mathbf{b}$ and $\mathbf{b} \times \mathbf{a}$ have the same length, $\mathbf{b} \times \mathbf{a}$ points in the opposite direction to that of $\mathbf{a} \times \mathbf{b}$, by the right-hand rule. Thus, $\mathbf{b} \times \mathbf{a} = -(\mathbf{a} \times \mathbf{b})$.

▼ Examples

- A nonzero cross product: $\langle 4, 5, 2 \rangle \times \langle 1, 3, 6 \rangle = \langle 24, -22, 7 \rangle$
- A zero cross product: $\langle 1, -2, 5 \rangle \times \langle -3, 6, -15 \rangle = \langle 0, 0, 0 \rangle$

▼ Group Work 1: Messing with the Cross Product

Four different problems are given for the group work. Give each group one problem to solve. If a group finishes early, give them another one to work out. After every group has finished their original problem, give them a few minutes to practice, and then have them come up and state their problem and demonstrate their solution.

Answers:

1. $\mathbf{c} \perp \mathbf{a}$ because $\mathbf{c}$ is the cross product of $\mathbf{a}$ and another vector. $\mathbf{c}$ is not perpendicular to $\mathbf{b}$. For example take $\mathbf{a} = \langle 1, 0, 0 \rangle$ and $\mathbf{b} = \langle 0, 1, 0 \rangle$.
2. Any vector of the form $\langle 0, t, 1 \rangle$ works.
3. The collection is a straight line; all vectors have the form $\left\langle \frac{4+k}{3}, \frac{2k-1}{3}, k \right\rangle$. (Other forms of this answer are acceptable.)
4. $\mathbf{a}$ is in the same plane as $\mathbf{a} - \mathbf{b}$ and in the same plane as $\mathbf{a} + \mathbf{b}$. Therefore, it is perpendicular to their cross product.

▼ Group Work 2: A Matter of Shading

This activity gives a nice application of cross products. Students will practice taking a wordy description of a problem, and converting it to some straightforward mathematics.

Answers:

Notice that, as suggested by the given illustration, the angles must always be between $90°$ and $180°$, so the normals must be chosen accordingly.

1. 2.265 radians or $129.76°$ **2.** 2.678 radians or $154.43°$ **3.** The second one

▼ Homework Problems

Core Exercises: 3, 6, 10, 14, 15, 23, 26, 31, 35

Sample Assignment: 3, 6, 7, 10, 12, 14, 15, 19, 23, 24, 26, 28, 31, 33, 35

GROUP WORK 1, SECTION 6.5

Messing with the Cross Product

1. Consider the vector $\mathbf{c} = \mathbf{a} \times (\mathbf{a} \times \mathbf{b})$. Is $\mathbf{c} \perp \mathbf{a}$? Is $\mathbf{c} \perp \mathbf{b}$? Either prove or give a counterexample.

2. Suppose that $\mathbf{j} \times \mathbf{a} = \mathbf{i}$. Give two possible solutions for $\mathbf{a}$, and discuss other possible solutions.

3. Given $\mathbf{a} = \langle 1, 2, 3 \rangle$ and $\mathbf{b} = \langle 1, -1, -1 \rangle$, sketch the collection of all position vectors $\mathbf{c}$ satisfying $\mathbf{a} \times \mathbf{b} = \mathbf{a} \times \mathbf{c}$.

4. Show that $\mathbf{a}$ is perpendicular to $(\mathbf{a} - \mathbf{b}) \times (\mathbf{a} + \mathbf{b})$.

GROUP WORK 2, SECTION 6.5

A Matter of Shading

One way to accurately render three-dimensional objects on a computer screen involves using the dot and cross products. In order to determine how to shade a piece of a surface we need to determine the angle at which rays from the light source hit the surface. To determine this angle, we compute the dot product of the light vector with the vector perpendicular to the surface at the particular point, called the normal vector.

If the light ray hits the surface straight on, that is, has an angle of 180° with the normal, then this piece of the surface will appear bright. On the other hand, if the light comes in on an angle, this piece of the surface will not appear as bright.

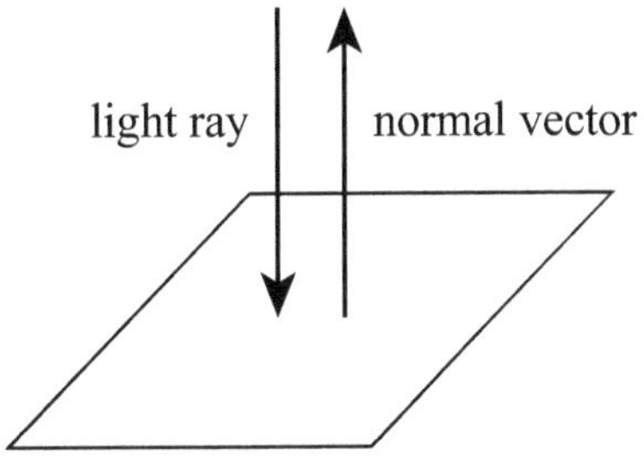

This piece appears bright.

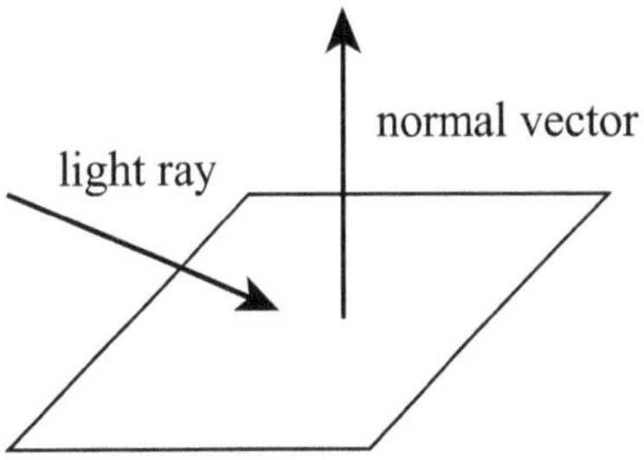

This piece appears dimmer.

1. Suppose the light source is placed directly above the xy-plane, so that the light rays come in parallel to the vector $\langle 0, 0, -1 \rangle$. At what angle (to the normal) do the light rays hit a triangle bounded by the points $(3, 2, 4)$, $(2, 5, 3)$, and $(1, 2, 6)$?
Note that your answer will be an obtuse angle (greater than 90°) because the vertex of the angle is formed by the tails of the two vectors.

2. Repeat Problem 1, this time using the region bounded by $(3, 5, 2)$, $(3, 3, 1)$, and $(1, 3, 1)$.

3. Suppose we are standing above the light source looking down on the xy-plane. Which of these two regions will appear brighter to us?

6.6 EQUATIONS OF LINES AND PLANES

▼ Suggested Time and Emphasis

1 class. Essential material.

▼ Points to Stress

1. Two ways to describe a line:
   - Vector equations (starting with point P_0 on the line and direction vector $\mathbf{d}$): $\mathbf{r} = \mathbf{r}_0 + t\mathbf{d}$.
   - Parametric equations: $x = x_0 + at$, $y = y_0 + bt$, $z = z_0 + ct$
2. Two ways to describe a plane:
   - Vector equation (starting with point P_0 and normal vector $\mathbf{n}$): $\mathbf{n} \cdot (\mathbf{r} - \mathbf{r}_0) = 0$.
   - Scalar equation: $a(x - x_0) + b(y - y_0) + c(z - z_0) = 0$ or $ax + by + cz + d = 0$.

▼ Sample Questions

- **Text Question:** When specifying the equation of a line in space, the text says that you need a point on the line and a vector parallel to the line. Why can't you determine a line in space simply by using one vector?
  Answer: One vector would give the direction of the line, but it could be placed anywhere in space. It would be analogous to trying to specify a line in $\mathbb{R}^2$ using only a slope.
- **Drill Question:** Find parametric equations of the line going through the points $(1, 2, 1)$ and $(-1, 3, 5)$.
  One Possible Answer: $x = 1 - 2t$, $y = 2 + t$, $z = 1 + 4t$
- **Drill Question:** A plane has normal vector $\mathbf{n} = \langle 1, 1, 0 \rangle$ and passes through the point $P(3, 1, 9)$. Find an equation for the plane.
  One Possible Answer: $x + y = 4$. Note that this is a plane perpendicular to the xy-plane, because z can be anything.

▼ In-Class Materials

- An overall theme for this section could be that a line is determined by a point and a direction, and a plane is determined by a point and a normal vector, or a point and two directions.
- Review parametric representation of lines in $\mathbb{R}^2$, and then generalize to $\mathbb{R}^3$. Recall that in two dimensions, a line can be determined by a point and a slope. Ask for the slope of the line between the points $(0, 0)$ and $(1, 2)$. Start with the line $y = mx + b$ and write it parametrically as $x = t, y = mt + b$. Then write the vector equation with $\mathbf{r}_0 = \langle 0, b \rangle$ and $\mathbf{d} = \langle 1, m \rangle$, a vector whose direction has slope m. Next ask for the slope of the line between $(0, 0, 0)$ and $(1, 2, 3)$. Note that there is no answer — we lose the idea of "slope" when going from two to three dimensions. So a vector is our only way of specifying direction in three dimensions.
- Discuss how to find the equation of a plane containing three non-collinear points P, Q, and R. We form the normal vector $\mathbf{n} = \mathbf{a} \times \mathbf{b}$, where $\mathbf{a} = \overrightarrow{PQ}$ and $\mathbf{b} = \overrightarrow{PR}$, and use the vector equation.
- Review lines in two dimensions. For example, ask the students to draw the line passing through $(1, 2)$ in the direction of $\mathbf{i} - 2\mathbf{j}$ and then write parametric equations for this line. Stress that these equations are not unique.

▼ Examples

- A plane that is relatively easy to sketch: $10x + 5y + 5z = 10$. (The intercepts are $x = 1$, $y = 2$, and $z = 2$.)

▼ Group Work 1: The Match Game

This is a pandemonium-inducing game. Each group is given three index cards. One has two points on a straight line, the second has parametric equations for a different line, and the last has a vector equation for a third line. The groups now are allowed to trade index cards with other groups. The first group to possess three descriptions for the same line wins a prize.

For the convenience of the teacher, each row below contains a winning combination. Make sure that each team starts with descriptions from different rows.

Category A	Category B	Category C
The line between $(0, 0, 1)$ and $(1, 2, 1)$	$\mathbf{r} = \langle 2, 4, 1\rangle + t\langle 1, 2, 0\rangle$	$x = 2t$ $y = 4t$ $z = 1$
The line between $(0, -3, 3)$ and $(3, 3, 0)$	$\mathbf{r} = \langle 1, -1, 2\rangle + t\langle 1, 2, -1\rangle$	$x = 2 + t$ $y = 1 + 2t$ $z = 1 - t$
The line between $(1, 3, 2)$ and $(1, -1, 6)$	$\mathbf{r} = \langle 1, 2, 3\rangle + t\langle 0, -1, 1\rangle$	$x = 1$ $y = -t$ $z = 5 + t$
The line between $(0, 0, 4)$ and $(12, 8, 8)$	$\mathbf{r} = \langle 9, 6, 7\rangle + t\langle -3, -2, -1\rangle$	$x = 6 - 6t$ $y = 4 - 4t$ $z = 6 - 2t$
The line between $(5, 0, 7)$ and $(-2, -7, 0)$	$\mathbf{r} = \langle 3, -2, 5\rangle + t\langle -1, -1, -1\rangle$	$x = 2 - 2t$ $y = -3 - 2t$ $z = 4 - 2t$
The line between $(-3, 3, -9)$ and $(3, -3, 9)$	$\mathbf{r} = \langle 0, 0, 0\rangle + t\langle -1, 1, -3\rangle$	$x = -1 + t$ $y = 1 - t$ $z = -3 + 3t$
The line between $(-4, 2, 1)$ and $(-11, 1, -1)$	$\mathbf{r} = \langle 3, 3, 3\rangle + t\langle 7, 1, 2\rangle$	$x = 10 + 7t$ $y = 4 + t$ $z = 5 + 2t$

▼ Group Work 2: Planes from Points

Give each group two sets of three points each, one noncollinear set and one collinear set. Ask the students to give an equation of the unique plane containing the points. For the second set of points this is a trick question, since collinear points do not determine a plane. Sample sets of points are given below.

$(0, 0, 0)$	$(-1, 4, 2)$	$(0, -5, 5)$	$(0, 0, 0)$	$(-1, 4, 2)$	$(0, -5, 5)$
$(1, 2, 3)$	$(3, 1, 1)$	$(0, 1, 1)$	$(1, 2, 3)$	$(3, 1, 1)$	$(0, 1, 1)$
$(2, 5, 9)$	$(7, 2, 0)$	$(0, 3, 4)$	$\left(\frac{3}{2}, 3, \frac{9}{2}\right)$	$(7, -2, 0)$	$(0, 3, 7)$

▼ Group Work 3: The Spanning Set

The purpose of this activity is to give the students a sense of how two non-parallel vectors in two dimensions span the entire xy-plane.

Start by giving each student or group of students a sheet of regular graph paper and a transparent grid of parallelograms formed by two vectors $\mathbf{u}$ and $\mathbf{v}$.

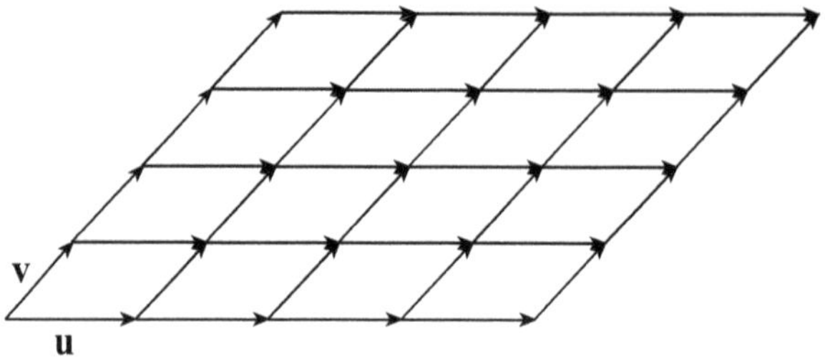

Next give each student a point (x, y) in the plane. By placing the grid over the graph paper they should estimate values of r and s such that $\langle x, y \rangle = r\mathbf{u} + s\mathbf{v}$. Repeat for several other points (including some for which one or both of r and s will be negative) until the students have convinced themselves that every point in the plane can be expressed in this manner.

Now repeat the activity with different vectors $\mathbf{u}$ and $\mathbf{v}$, perhaps using the same points as before.

As a wrap-up, give the students specific vectors $\mathbf{u}$ and $\mathbf{v}$, such as $\mathbf{u} = \langle 3, 1 \rangle$ and $\mathbf{v} = \langle -1, -2 \rangle$, and have them determine values of r and s for several points. See if they can find general formulas for r and s in terms of the point (x, y). What goes wrong algebraically if $\mathbf{u}$ and $\mathbf{v}$ are parallel?

▼ Homework Problems

Core Exercises: 5, 6, 11, 16, 18, 21, 24, 32

Sample Assignment: 5, 6, 8, 11, 14, 16, 18, 20, 21, 24, 25, 27, 32, 34, 36

GROUP WORK 1, SECTION 6.6

The Match Game

Your teacher has just handed you three index cards, each with an equation of a different line. Your task is, by clever trading with other groups, to wind up with three different descriptions of the *same line.* The winning team gets a boffo prize, so go for it!

GROUP WORK 2, SECTION 6.6

Planes from Points

1. Consider the following set of three points:

POINT 1: __________
POINT 2: __________
POINT 3: __________

Find an equation of the unique plane containing these points.

2. Repeat Part 1 using these points:

POINT 1: __________
POINT 2: __________
POINT 3: __________

The Spanning Set

7 CONIC SECTIONS

7.1 PARABOLAS

▼ Suggested Time and Emphasis

$\frac{1}{2}$ class. Recommended material.

▼ Points to Stress

1. The definition and geometry of parabolas.
2. Using the equation of a parabola to find relevant constants.
3. Graphing a parabola given its equation.

▼ Sample Questions

- **Text Question:** How can you tell if the axis of a parabola is vertical or horizontal?

  Answer: If there is a y^2, term it is horizontal; if there is an x^2 term, it is vertical.
- **Drill Question:** Find the focus and directrix of $y = \frac{1}{16}x^2$ and sketch its graph.

  Answer: Focus $\left(0, \frac{1}{4}\right)$, directrix $y = -\frac{1}{4}$

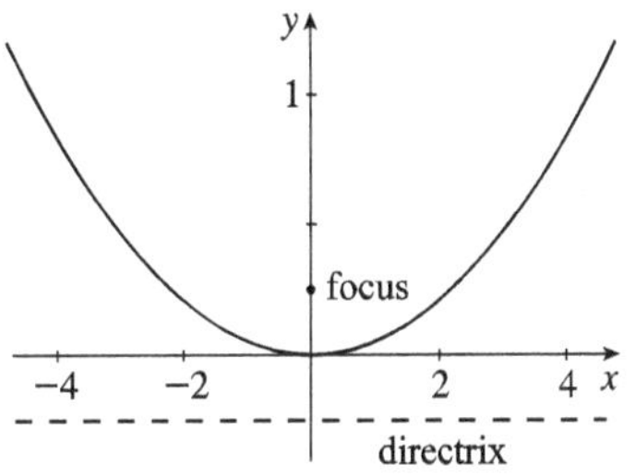

▼ In-Class Materials

- Have students sketch a parabola from scratch. Hand out a sheet of paper with focus and directrix, and hand out rulers. Get the students to plot the points from which the distance to the directrix is equal to the distance to the focus. Have them keep plotting points until a parabolic shape emerges.
- Using dental floss and modeling compound (such as clay or Play-Doh) it is easy for the students to make half a cone and slice it. Have the class attempt to do so to get a parabola. Most of them will wind up creating half of a hyperbola. Refer these students to the chapter overview. Make sure that they note a parabola is not as easy to create this way as a hyperbola or an ellipse. If the cut is at the wrong angle, even slightly, a hyperbola or an ellipse will be formed instead of a parabola.
- Discuss the reflection properties of parabolas: A beam of light originating at the origin will emerge parallel to the parabola's axis of symmetry, and a beam of light that is parallel to the parabola's axis of symmetry will reflect off of the parabola in a direction that goes through its focus. One nice project is to construct a parabolic pool table or miniature golf hole. The students accurately graph a parabola, and glue erasers or wood along its border, placing the "hole" at the focus. A golf ball or pool ball that is rolled in a direction parallel to the axis will always bounce into the hole.

- As of this writing, it is possible to purchase a parabolic listening device for about \$40 on eBay, or about \$60 new. If this is feasible, many experiments and demonstrations can be done. For example, students can whisper from a long distance away and be heard using the device.
- Make the connection between quadratic functions and parabolas explicit. Point out that any equation of the form $y = ax^2 + bx + c$ can be written as $y = n(x-h)^2 + k$ by completing the square, yielding a (possibly shifted) parabola. The constant n is often written as $4p$. Note that if $a = 0$ we have a line. So we can call a line a "degenerate parabola".

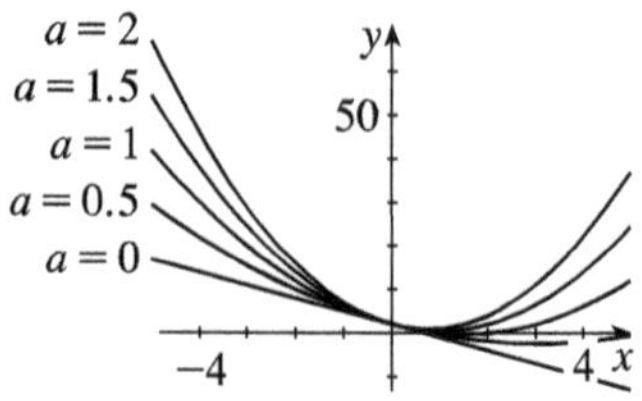

▼ Examples

A shifted parabola:

$$\begin{aligned} y &= 2x^2 - 8x + 2 \\ &= 2\left(x^2 - 4x\right) + 2 \\ &= 2\left(x^2 - 4x + 4\right) - 6 \\ &= 2(x-2)^2 - 6 \end{aligned}$$

$p = \frac{1}{2}$, focus $\left(0+2, \frac{1}{2} - 6\right) = \left(2, -\frac{11}{2}\right)$, directrix $y = -\frac{1}{2} - 6 = -\frac{13}{2}$

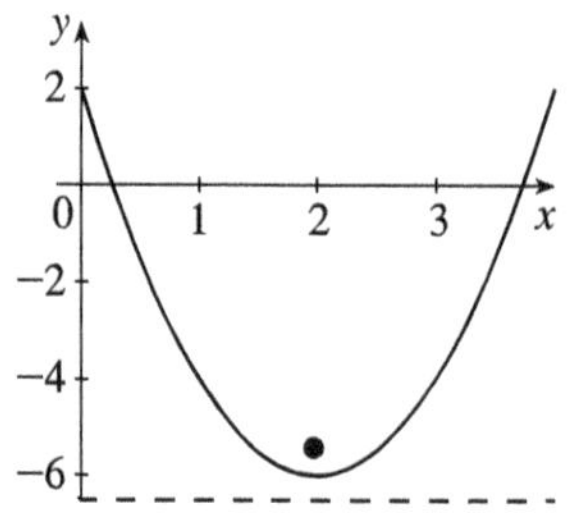

▼ Group Work: The Effect of Gravity

Demonstrate that a tossed object will trace out a parabola. A quick way to do this is, of course, to throw an eraser. A more elaborate way to do it is to darken the room and toss a glow-stick or tiny flashlight. Starting with Problem 5, this activity segues into shifting the parabola, which will be covered in more detail in Section 7.4.

Answers:

1.

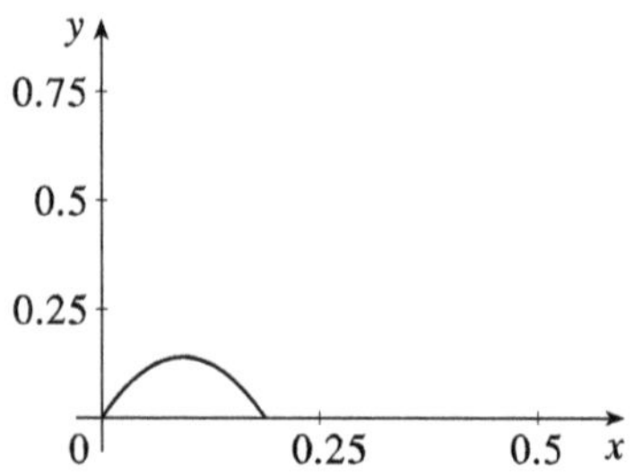

2.

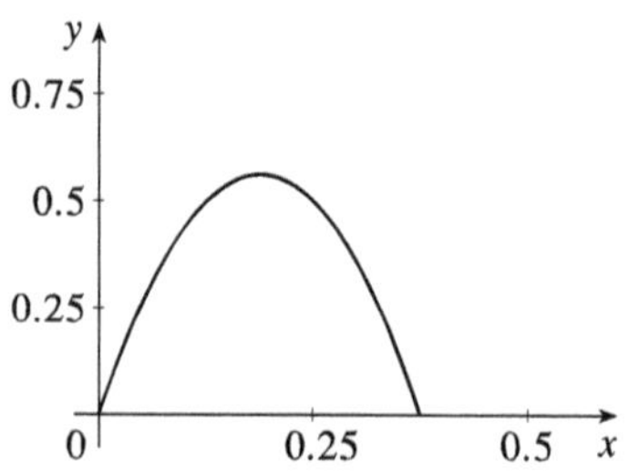

3. The second: it has a higher initial velocity, so its maximum height is greater.

4. The second: it has a higher initial velocity, so it travels upward longer.

5. $h(t) = -16t^2 + v_0 t + 100$

6. $h(t) = -16t^2 + v_0 t + s_0$

7. (a) $0 = -16t^2 + 216 \quad \Rightarrow \quad t \approx 3.674$ s

(b) $0 = -16t^2 + 10t + 216 \quad \Rightarrow \quad t = 4$ s

(c) $h(t) = -16t^2 + 10t + 216 = -16\left(t^2 - \frac{5}{8}t + \frac{25}{256}\right) + 216 + \frac{25}{256} = -16\left(t - \frac{5}{16}\right)^2 + \frac{55{,}321}{256}$. It reaches its maximum height at $t = \frac{5}{16}$ s.

(d) The maximum height is $\frac{3481}{16} \approx 217.56$ ft.

▼ Homework Problems

Core Exercises: 3, 7, 14, 18, 31, 40, 42, 53

Sample Assignment: 3, 7, 10, 14, 18, 21, 31, 36, 40, 42, 46, 49, 53, 55, 56

GROUP WORK, SECTION 7.1

The Effect of Gravity

If you toss an object upward from the ground, its height at time t (in feet) is given by

$$h(t) = -16t^2 + v_0 t$$

where v_0 (pronounced "vee-naught") is the velocity of the toss in ft/s.

1. Sketch the graph of the height of a ball as a function of time if it is thrown with an initial velocity of 3 ft/s.

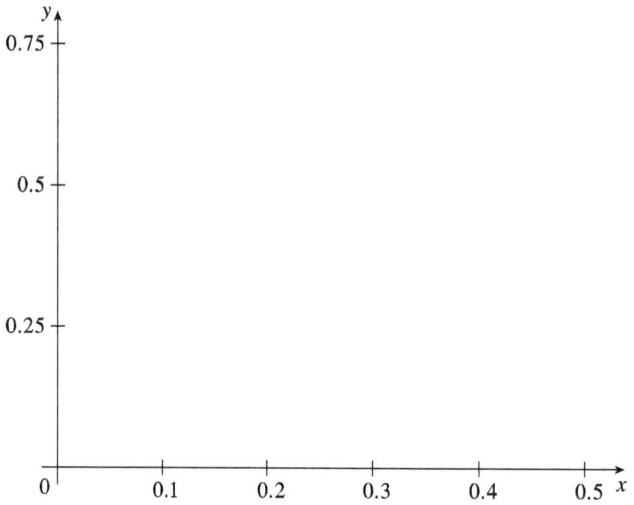

2. Sketch the graph of the height of a ball if it is thrown with an initial velocity of six feet per second.

3. Which of your two graphs has a higher vertex? Physically, why is that the case?

4. Which of your two graphs has a vertex that is farther to the right? Physically, why is that the case?

5. Now assume you are not starting at the ground, but at a height of 100 feet. Find a formula for the height of the object as a function of time.

6. Find a formula for the height of the object if you are starting s_0 feet above the ground.

7. (a) Assume now that you take a ball to the top of a 216 foot tall building and let it drop. When does it reach the ground?

(b) When would it reach the ground if you tossed it upwards with an initial velocity of 10 ft/s?

(c) In that case, when would it reach its maximum height?

(d) What would that maximum height be?

7.2 ELLIPSES

▼ Suggested Time and Emphasis

$\frac{1}{2}$–1 class. Optional material.

▼ Points to Stress

1. The definition and geometry of ellipses.
2. Using the equation of an ellipse to find relevant constants and to graph the ellipse.
3. Eccentricity.

▼ Sample Questions

- **Text Question:** Does $\frac{x^2}{5^2} + \frac{y^2}{6^2} = 1$ describe a horizontal or a vertical ellipse? How do you know?
  Answer: It is vertical because the denominator of the y^2 term is larger than that of the x^2 term.
- **Drill Question:** Find the vertices and foci of the ellipse $\frac{1}{4}x^2 + y^2 = 1$ and sketch its graph.
  Answer: Vertices $(\pm 2, 0)$, foci $(\pm\sqrt{3}, 0)$

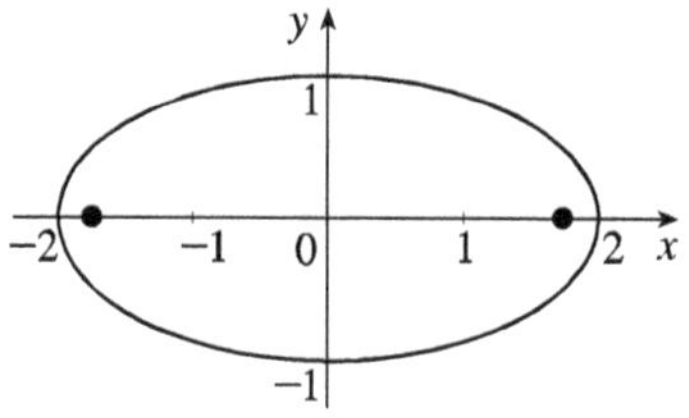

▼ In-Class Materials

- Have students sketch an ellipse with thumbtacks and string, as suggested in the text. Have some use foci that are close together, and some that are farther apart. Make sure they see the connection between this activity and the idea that the summed distance from the foci is a constant.
- As noted in Section 7.1, one can use dental floss and a modeling compound (such as clay or Play-Doh) to make half a cone and slice it. Have the class attempt to do so to get a circle. Note that if their angle is slightly off, they will get an ellipse. Make the analogy that just as a square is a particular kind of rectangle, a circle is a particular kind of ellipse.
- Many representational artists never draw circles, noting that it is rare in nature to see a circle, since we are usually looking at an angle, thus seeing an ellipse. Perhaps have the students bring in photographs of manhole covers and other "circular" objects, noting that if the camera angle is not straight on, the resultant image is elliptical. (Because of the optical illusion of perspective, it makes things easier to draw the outline

of the "circle" with a marker to see that it is an ellipse.)

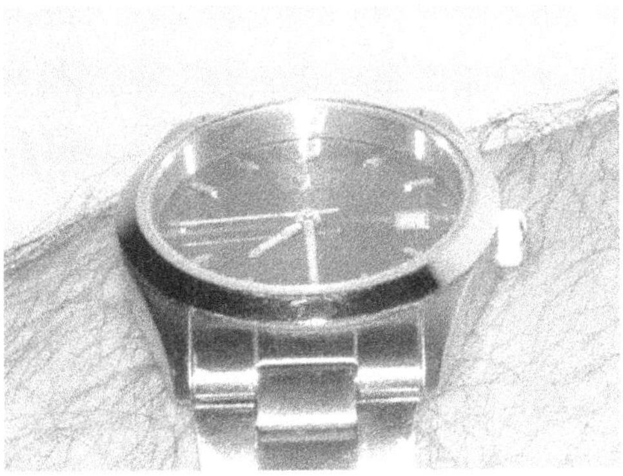

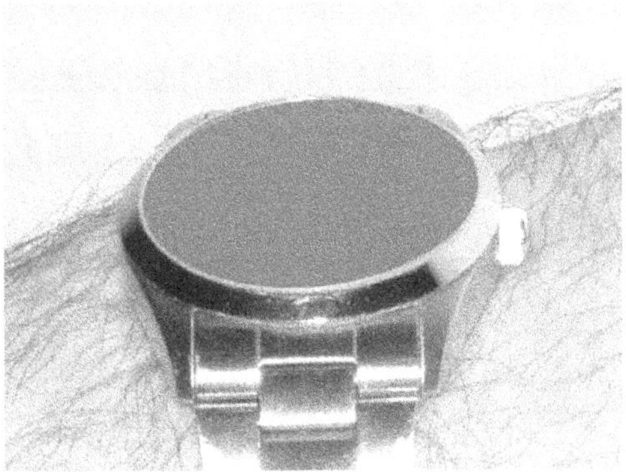

- Discuss the reflection property of an ellipse: a beam of light originating at one focus will reflect off the ellipse and pass through the other focus. One nice project is to construct an elliptical pool table. The students accurately graph an ellipse, and glue erasers or wood along its border, placing the "hole" at one focus and marking the second. A golf ball or pool ball that is placed on the mark and struck in any direction will ricochet into the hole.
- Note that if $a = b$ then we have only one focus, at $(0, 0)$. In this case, the geometric definition breaks down, but it is clear from the equation that we have a circle. This is why a circle can be thought of as a "degenerate ellipse".

▼ Examples

- A vertical ellipse: $\dfrac{x^2}{16} + \dfrac{y^2}{25} = 1$

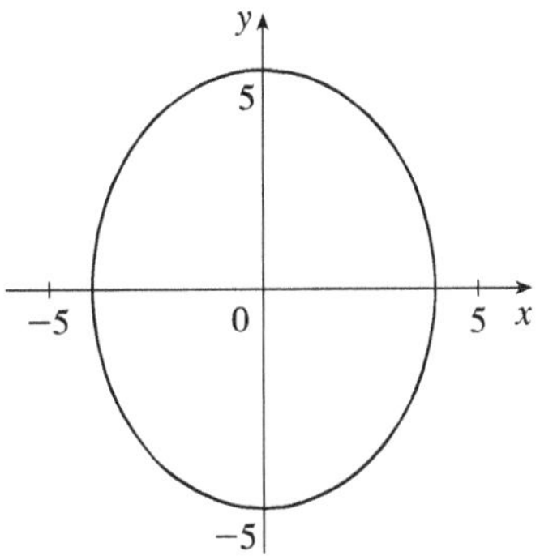

- A horizontal ellipse: $\dfrac{x^2}{9} + 4y^2 = 1$

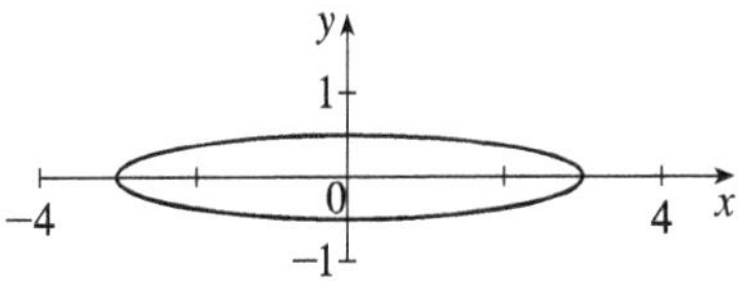

▼ Group Work: Ellipses in Nature

As suggested earlier, have the students bring in photographs of circles, either clipped out of magazines or taken by themselves. Have them use a photocopier to make the photograph sufficiently large. Using a photocopier or a transparency, instruct the students to overlay a grid on their photograph, and determine whether the shape in question is a circle or an ellipse. Have the students find the equation of their shape.

▼ Homework Problems

Core Exercises: 2, 6, 9, 20, 28, 33, 40, 48, 55

Sample Assignment: 2, 6, 8, 9, 13, 20, 24, 28, 33, 36, 40, 43, 48, 51, 54, 55

7.3 HYPERBOLAS

▼ Suggested Time and Emphasis

$\frac{1}{2}$ class. Optional material.

▼ Points to Stress

1. The definition and geometry of hyperbolas.
2. Using the equation of a hyperbola to find relevant constants.
3. Graphing a hyperbola given its equation.

▼ Sample Questions

- **Text Question:** Is $\frac{x^2}{5^2} - \frac{y^2}{6^2} = 1$ the equation of a horizontal or a vertical hyperbola? How do you know?
  Answer: It is horizontal because the x^2 term is positive.
- **Drill Question:** Find the vertices, foci, and asymptotes of the hyperbola given by $\frac{y^2}{4} - x^2 = 1$. Then graph it.
  Answer: Vertices $(0, \pm 2)$, foci $(0, \pm\sqrt{5})$, asymptotes $y = \pm 2x$

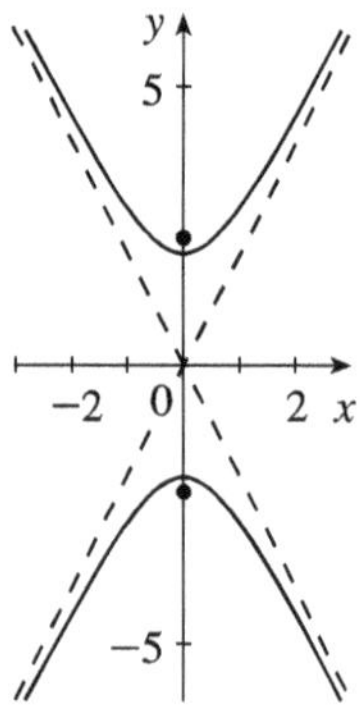

▼ In-Class Materials

- Have students sketch a hyperbola "from scratch." Hand out a sheet of paper with two foci, and hand out rulers. Get the students to plot points where the difference of the distances between the points and the foci is 1 inch. Have them keep plotting points until a hyperbolic shape emerges. There is another way to sketch a hyperbola which will be discussed in the Group Work.
- As noted in Section 7.1, one can use dental floss and a modeling compound (such as clay or Play-Doh) to make a half-cone and slice it. Have the students attempt to do so to get a hyperbola. Notice that the slice does not have to be straight up and down, as shown in the prologue to the chapter. As long as the slice would cut the other half of the cone, the resultant curve is a hyperbola.
- Discuss the reflection property of hyperbolas: Take a point between the branches, and aim a beam of light at one of the foci. It will reflect off the hyperbola, and go in a path aimed directly at the other focus. This reflection property is harder to model physically than those of the ellipse and the parabola.

▼ Examples

- A vertical hyperbola: $-\dfrac{x^2}{16}+\dfrac{y^2}{25}=1$

  Vertices $(0,\pm 5)$, asymptotes $y=\pm\frac{5}{4}x$, foci $\left(0,\pm\sqrt{41}\right)$

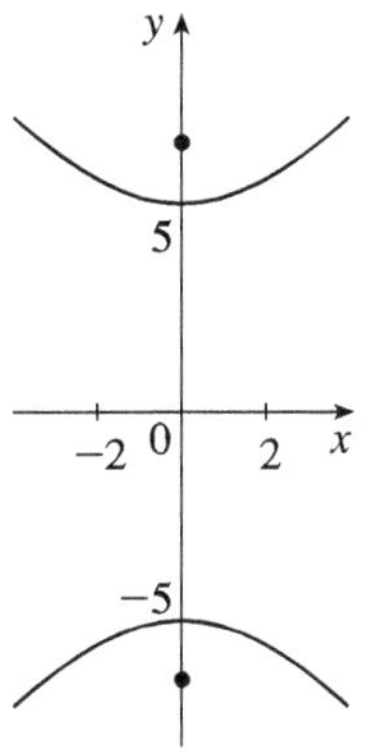

- A horizontal hyperbola: $\dfrac{x^2}{9}-2y^2=1$

  Vertices $(\pm 3,0)$, asymptotes $y=\pm\frac{1}{3\sqrt{2}}x$, foci $\left(\pm\sqrt{\frac{37}{2}},0\right)$

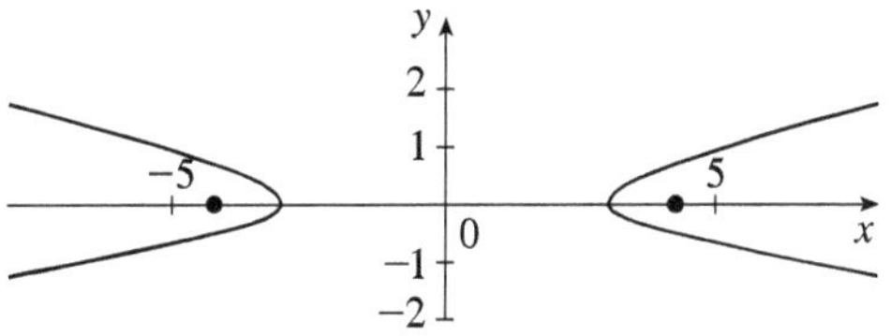

▼ Group Work: Sketching Hyperbolas

This will require students to follow directions carefully. You may have to model the process for them on an overhead projector. There is a bit of a "knack" to keeping the string in the correct position. The students will need a cardboard ruler or straightedge, a thumbtack, tape or fast-drying glue, string, scissors, a pencil, and patience.

If students finish early, there is another construction they can try: Mark two points A and B on a piece of paper. Draw a circle centered at A with a radius small enough so that B lies outside the circle. Now mark an arbitrary point on the circle and fold the paper so B touches that point. Crease the paper along the fold. Mark a different point on the circle, and crease again. If you do this enough times, the creases will form a filled-in hyperbola. As of this writing, the result of this process can be seen on the web at `http://www.cognitiohk.edu.hk/maths/math/em03con.htm`.

Answer:

Let the length of the string be s and the length of the ruler r. Call the unlabeled end of the ruler C. Consider a point on your curve — call it P. (In the figure, the tip of the pencil is at P.)

Then

$$\begin{aligned} \left|\overline{BP} - \overline{AP}\right| &= \left|\overline{BP} - \left(s - \overline{CP}\right)\right| \\ &= \left|\overline{BP} + \overline{CP} - s\right| \\ &= |r - s| \end{aligned}$$

Thus $\overline{BP} - \overline{AP}$ is a constant, and so by definition the collection of points P constitutes a hyperbola.

▼ Homework Problems

Core Exercises: 2, 5, 15, 22, 32, 41, 46, 49

Sample Assignment: 2, 5, 8, 15, 20, 22, 25, 32, 36, 41, 46, 48, 49

GROUP WORK, SECTION 7.3

Sketching Hyperbolas

We are going to be constructing the graph of a hyperbola. This method is going to require some patience, and mastering an unnatural motion, but it does work!

1. Mark two points A and B on a piece of paper. They will be the foci of your hyperbola.
2. Cut a piece of string, a bit longer or a bit shorter than your ruler.
3. Attach one end of the string to one end of your ruler, and the other end of the string to A. Attach it firmly—you don't want it to move about.
4. Pin the other end of the ruler to B, so that the ruler is free to rotate around B.
5. Now tighten the string with a pencil, as shown in the diagram, and rotate the ruler. Make sure that the string does not stretch.

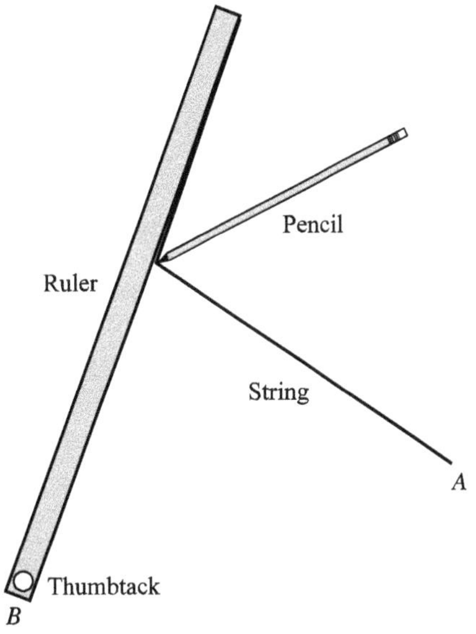

6. To get the other half of the hyperbola, perform the same trick with the end of the ruler tacked to point A and the end of the string attached to point B.

Congratulations! You are now the proud parent of a lovely little hyperbola. Now figure out why this process works.

7.4 SHIFTED CONICS

▼ Suggested Time and Emphasis

1 class. Optional material. May be combined with Sections 7.1–7.3.

▼ Points to Stress

1. Completing the square in a general equation of a conic in order to apply the techniques of Section 2.2 to graph the conic.
2. Identifying conic sections by the constants in the general equation.
3. Understanding degenerate conic sections.

▼ Sample Questions

- **Text Question:** Identify the conic section with equation $9x^2 - 36x + 4y^2 = 0$.
  Answer: Ellipse
- **Drill Question:** Graph the conic section with equation $4x^2 - 16x + 9y^2 - 20 = 0$.
  Answer: $4x^2 - 16x + 9y^2 - 20 = 0 \quad \Leftrightarrow \quad 4(x-2)^2 + 9y^2 = 36 \quad \Leftrightarrow \quad \dfrac{(x-2)^2}{9} + \dfrac{y^2}{4} = 1$

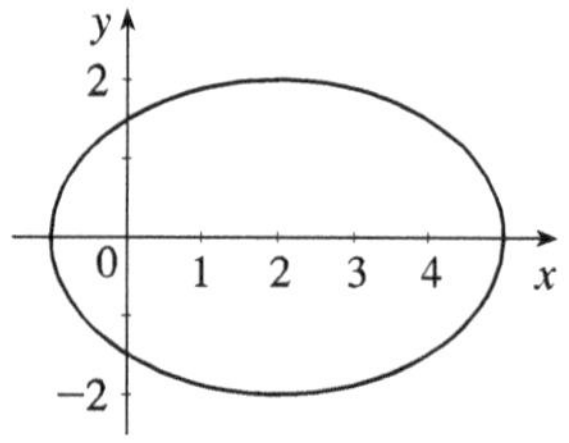

▼ In-Class Materials

- Class time can be saved if this section is covered along with Sections 7.1–7.3. For example, after discussing parabolas, immediately do an example of a shifted parabola. Teach ellipses and hyperbolas similarly.
- The text's guide to the general equation of a shifted conic is important, and students should not memorize it blindly. Point out that if they've learned the previous section, these rules of thumb should make perfect sense. Have them look at the three general formulas they have learned, generalizing them to allow for a shift:

$$y = 4p(x-h)^2$$

$$\frac{(x-h)^2}{a^2} + \frac{(y-k)^2}{b^2} = 1$$

$$\frac{(x-h)^2}{a^2} - \frac{(y-k)^2}{b^2} = 1$$

- Help the class recognize the difference between the formulas of conic sections by pointing out that the equation of a parabola has degree 2 in only one variable (which variable it is determines which type of parabola it is). Equations of ellipses and hyperbolas have degree 2 in both variables; for ellipses the second-degree terms have the same sign, and for hyperbolas they have opposite signs.

▼ Examples

- A nondegenerate ellipse:

$$25x^2 + 4y^2 - 150x + 40y + 324 = 0 \quad \Leftrightarrow$$
$$\frac{(x-3)^2}{4} + \frac{(y+5)^2}{25} = 1$$

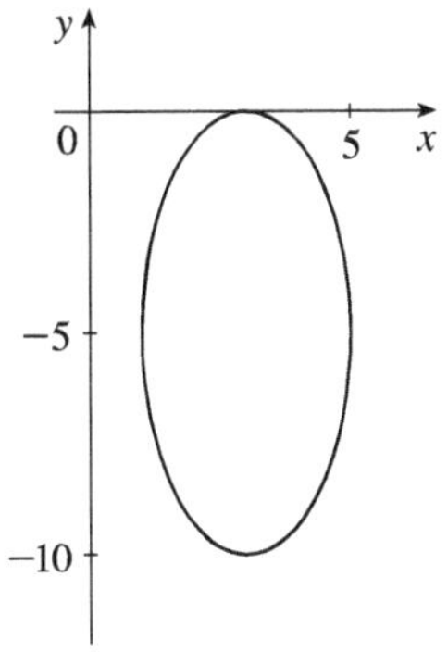

- A degenerate hyperbola:

$$x^2 - y^2 - 4x - 2y + 3 = 0 \quad \Leftrightarrow$$
$$(x-2)^2 - (y+1)^2 = 0$$

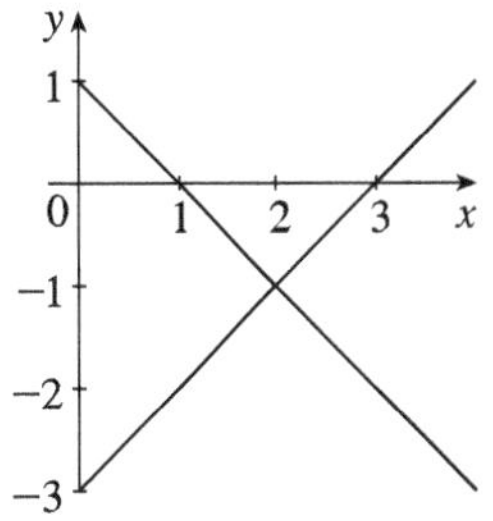

▼ Group Work: It's Easier in Polar

This activity is for classes which have covered polar coordinates.

Answer: The graph is an ellipse if $e < 1$, a hyperbola if $e > 1$, and a parabola when $e = 1$. The directrix is $x = \pm d$ or $y = \pm d$.

▼ Homework Problems

Core Exercises: 5, 10, 18, 24, 27, 42, 40

Sample Assignment: 5, 8, 10, 11, 18, 21, 24, 27, 34, 40, 42, 43

GROUP WORK, SECTION 7.4

It's Easier in Polar

Consider the polar graph with equation

$$r = \frac{ed}{1 \pm e\sin\theta}$$

It turns out this is always the graph of a (possibly degenerate) conic section. Play with the positive constants d and e. Under what circumstances is the graph an ellipse? A parabola? A hyperbola?

7.5 ROTATION OF AXES

▼ Suggested Time and Emphasis

1 class. Optional material.

▼ Points to Stress

1. Rotation of axes formulas.
2. Graphing a general quadratic equation, including middle terms.

▼ Sample Questions

- **Text Question:** If you are presented with the equation of a conic such as $3x^2 + 5xy - 2y^2 + x - y + 4 = 0$, how would you go about finding if it is an ellipse, parabola, or hyperbola?

  Answer: Evaluate the discriminant.
- **Drill Questions:** Consider the equation $3x^2 + 5xy - 2y^2 + x - y + 4 = 0$. What shape is its graph?

  Answer: It is a hyperbola.

▼ In-Class Materials

- Point out the progression of the past few sections. Write out $Ax^2 + Bxy + Cy^2 + Dx + Ey + F = 0$. When the course started, we studied equations of lines, which was just the case in which $A = B = C = 0$. When we discussed circles, that was the case where $A = C$, $B = D = E = 0$. Shifting the center of a circle allowed D and E to take on nonzero values. Our study of conic sections removed the restriction that $A = C$, so this section really is the last step in a logical progression.
- This is a good section to teach with the aid of a CAS, or even a good place to introduce such a device. A great deal of the work in graphing an arbitrary conic section is multiplying out long algebraic expressions, which a CAS can do easily.
- If the students are going to be moving on to linear algebra in the future, Exercise 37 is an excellent one to go over in class. It addresses the idea of transformations, which will be a major topic in that course.

▼ Examples

- The rotated parabola $x^2 + 2xy + y^2 + 2x - 4y = 0$:

  $\cot(2\phi) = 0 \Rightarrow 2\phi = \frac{\pi}{2} \Rightarrow \phi = \frac{\pi}{4}$. Now we have $x = \sqrt{2}X - \sqrt{2}Y$, $y = \sqrt{2}X + \sqrt{2}Y$. After substituting, and a lot of algebra, we get $8x^2 - 2\sqrt{2}x - 6\sqrt{2}y = 0$. Completing the square, $y = \frac{2\sqrt{2}}{3}\left(x - \frac{\sqrt{2}}{8}\right)^2 - \frac{\sqrt{2}}{48} \approx 0.9\,(x - 0.2)^2$ (for sketching purposes).

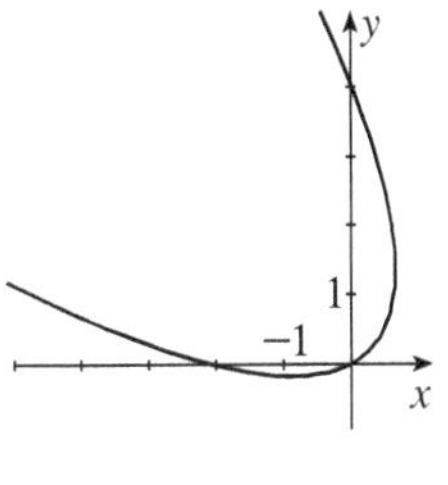

▼ Group Work

Graphing any of the conics in Exercises 15–28 makes a good group project, but be aware that these problems will take the students a significant amount of time.

▼ Homework Problems

Core Exercises: 4, 10, 18, 25, 28, 31, 35

Sample Assignment: 4, 8, 10, 14, 18, 21, 25, 28, 31, 34, 35, 36, 37

7.6 POLAR EQUATIONS OF CONICS

▼ Suggested Time and Emphasis

1 class. Optional material.

▼ Points to Stress

1. Expressing conic sections in polar form.
2. Rotating conic sections in parametric form.
3. The eccentricity of a conic.

▼ Sample Questions

- **Text Question:** State a reason that we would want to express a conic in polar form.

  Answer: Answers will vary. One reason is that they are easier to rotate when expressed in polar form.

- **Drill Question:** Identify the conic $r = \dfrac{6}{1 \pm 3\cos\theta}$

  Answer: It is a hyperbola.

▼ In-Class Materials

- It is good to go over Figure 6 with the class, having them use their calculators to vary the eccentricity e of a given conic section, and see how it affects the graph. Most graphic calculators have a mode for graphing polar functions as well.
- Notice the advantages of working with conics in polar form. They are easier to rotate, for example. It is also easier to identify a conic as an ellipse, parabola, or hyperbola by inspection. After some practice, we can even look at such an equation and have a good idea of what the graph will look like, again by inspection.
- We have called a circle a "degenerate ellipse". The text notes that when e is close to zero, the resultant ellipse is close to circular. Examine the equation of a conic when $e = 0$ to see how it is undefined for that value of e.

▼ Examples

Identifying and graphing conic sections:

- $r = \dfrac{6}{1 + 3\cos\left(\theta - \frac{\pi}{6}\right)}$ is a hyperbola, rotated 30° from standard position, with eccentricity 3.

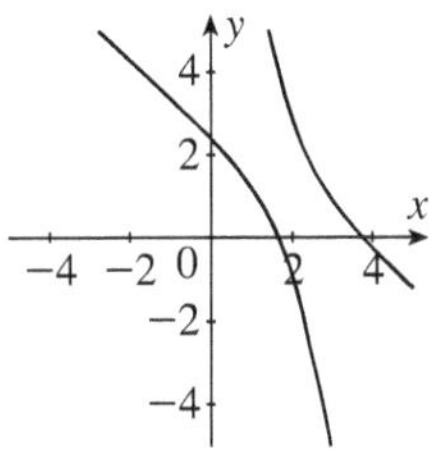

- $r = \dfrac{6}{3+3\cos\left(\theta - \frac{\pi}{6}\right)} = \dfrac{2}{1+\cos\left(\theta - \frac{\pi}{6}\right)}$ is a parabola.

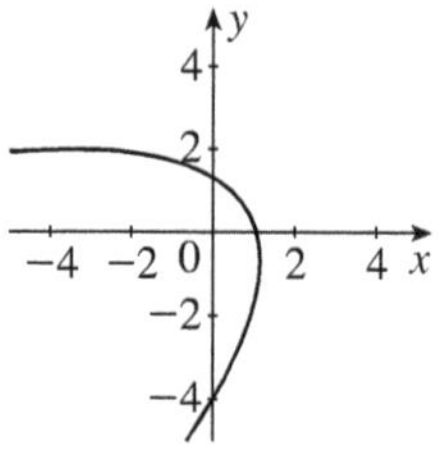

- $r = \dfrac{6}{6+3\cos\left(\theta - \frac{\pi}{6}\right)} = \dfrac{1}{1+\frac{1}{2}\cos\left(\theta - \frac{\pi}{6}\right)}$ is an ellipse.

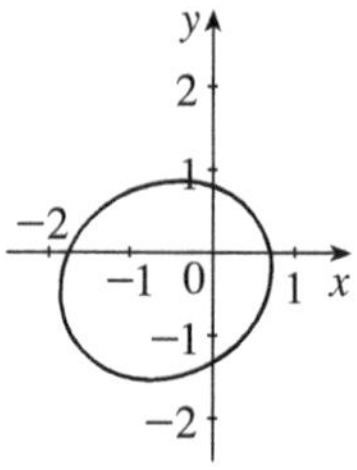

▼ Group Work: The Polar Conic

This activity allows students to discover the last topic in the section, and reinforces the formulas in the section.

Answers:

1. For $e > 1$ we have a hyperbola, for $e = 1$ we have a parabola, and for $0 < e < 1$ we have an ellipse. Also, as $e \to 0$, the ellipse becomes more circular.
2. As c changes, the ellipse rotates about the origin.
3. Technically, we are moving the directrix. The students should notice that the size of the ellipse changes.

▼ Homework Problems

Core Exercises: 3, 7, 12, 18, 23, 26, 32, 43

Sample Assignment: 3, 7, 9, 12, 14, 18, 20, 23, 26, 27, 32, 33, 38, 43, 44

GROUP WORK, SECTION 7.6

The Polar Conic

1. Consider the polar equation $r = \dfrac{2e}{1 + e\sin\theta}$. Graph this equation for $e = 4, 3, 2, 1.5, 1, 0.8, \frac{1}{2}, \frac{1}{3}, \frac{1}{4}$, and $\frac{1}{5}$. What do you notice?

2. Now consider the equation $r = \dfrac{1}{1 + \frac{1}{2}\sin(\theta + c)}$. Graph this equation for various values of c, such as $c = -0.3, -0.2, -0.1, 0, 0.1, 0.2$, and 0.3. What do you notice?

3. Finally, consider the equation $r = \dfrac{\frac{1}{2}d}{1 + \frac{1}{2}\sin\theta}$. Graph this equation for various values of d. What do you notice?

8 EXPONENTIAL AND LOGARITHMIC FUNCTIONS

8.1 EXPONENTIAL FUNCTIONS

▼ Suggested Time and Emphasis

1 class. Essential material.

▼ Points to Stress

1. The definition of an exponential function, including what it means to raise a to an irrational number.
2. The geometry of exponential functions and their transformations.
3. Periodically compounded interest.

▼ Sample Questions

- **Text Question:** Sketch a graph of $y = \left(\frac{1}{2}\right)^x$.

  Answer:

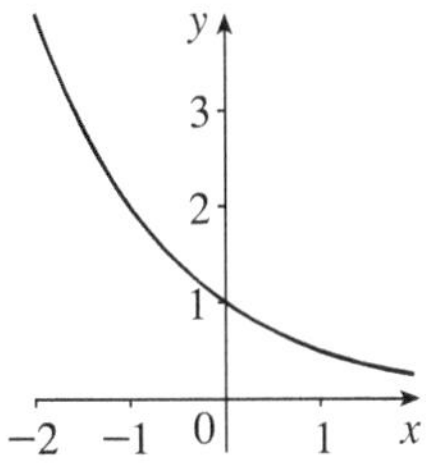

- **Drill Question:** Find the total amount of money in an account after 2 years if $100 is invested at an interest rate of 5.5% per year, compounded quarterly.

  Answer: $\$111.54$

▼ In-Class Materials

- Start to draw a graph of $y = 2^x$, using a carefully measured scale of one inch per unit on both axes. Point out that after one foot, the height would be over 100 yards (the length of a football field). After two feet, the height would be 264 miles, after three feet it would be 1,000,000 miles (four times the distance to the moon), after three and a half feet it would be in the heart of the sun. If the graph extended five feet to the right, $x = 60$, then y would be over one light year up.
- Point out this contrast between exponential and linear functions: For equally spaced x-values, linear functions have constant *differences* in y-values, whereas pure exponential functions have constant *ratios* in y-values. Use this fact to show that the following table describes an exponential function, not a linear one.

x	y
-6.2	0.62000
-2.4	0.65100
1.4	0.68355
5.2	0.71773
9.0	0.75361
12.8	0.79129

- Estimate where $3^x > x^3$ and where $2^x > x^8$ using technology. Notice that exponential functions start by growing much more *slowly* than polynomial functions, and then wind up growing much more *quickly*. For example, if one were to graph x^2 versus x using one inch per unit, then when $x = 60$, y would be only 100 yards, as opposed to a light year for $y = 2^x$. (The sun is only 8 light minutes from the earth.)

▼ Examples

- A shifted exponential curve:

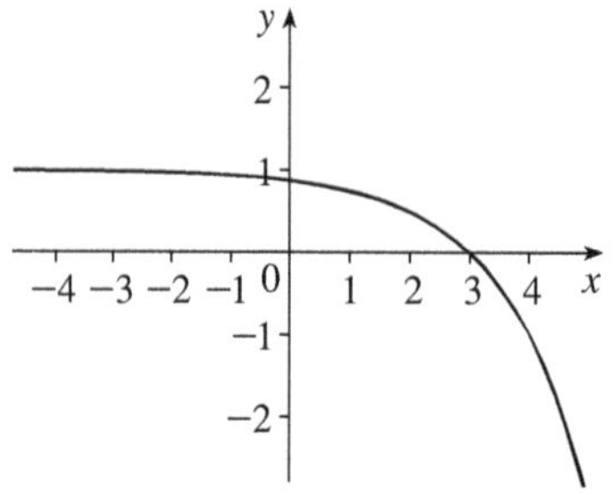

$$f(x) = -2^{x-3} + 1$$

- A comparison of compounding rates: \$2000 is put in an IRA that earns 7%. Its worth after 10 years is given in the following table.

Compounded annually	\$3934.30
Compounded semi-annually	\$3979.58
Compounded monthly	\$4019.32
Compounded weekly	\$4025.61
Compounded daily	\$4027.24
Compounded hourly	\$4027.50

Note: Most calculators will not be able to determine the amount earned if the interest is compounded, say, every tenth of a second. Underflow errors in the microprocessor will make the computation difficult.

▼ Group Work 1: I've Grown Accustomed to Your Growth

Before handing out this activity, it may be prudent to review the rules of exponentiation. This exercise enables students to discover for themselves the equal ratio property of exponential functions.

Answers:

1. Yes ($m = 1$), no, yes ($m \approx 2.08$), yes ($m \approx 2.01$)
2. Equally spaced changes in x-values result in equally spaced changes in y-values
3. Equally spaced changes in x values result in equally proportioned changes in y-values with the same ratio. $b = 2$, $b = 0.9975$, $b = 2.25$, $b = 3$
4. The "$+ C$" gets in the way when taking the ratio. However, the property is close to being true when A and b are large compared to C.

▼ Group Work 2: Comparisons

The purpose of this group work is to give students a bit of "picture sense". It is acceptable if they do this by looking at the graphs on their calculators, setting the windows appropriately.

Answers:

1. $0 < x < 1.374$ and $x > 9.940$

2. $0 < x < 1.051$ and $x > 95.7169$

3. $0 < x < 1.17$ and $x > 22.5$

4. $0 < x < 1.34$ and $x > 10.9$

5. $0 < x < 1.077$ and $x > 58.7702$

▼ Homework Problems

Core Exercises: 6, 9, 12, 21, 28, 42, 47, 52, 55

Sample Assignment: 6, 8, 9, 12, 18, 21, 23, 28, 35, 38, 42, 47, 52, 55, 59

GROUP WORK 1, SECTION 8.1

I've Grown Accustomed to Your Growth

1. Two or three of the following four tables of data have something in common: linear growth. Without trying to find complete equations of lines, determine which of them are linear growth, and determine their rate of change:

x	y
1	2
2	3
3	4
4	5

x	y
21.5	4.32
32.6	4.203
43.7	4.090
54.8	3.980

x	y
−3	1.1
−2.5	2.14
−2	3.18
−1.5	4.25

x	y
1	−5.00
3	−0.98
6	5.05
8	9.07

2. In a sentence, describe a property of linear growth that can be determined from a table of values.

3. The following four tables of data have something in common: exponential growth. Functions of the form $y = Ab^x$ (or Ae^{kx}) have a property in common analogous to the one you stated in Question 2. Find the property, and then find the value of b.

x	y
1	5
2	10
3	20
4	40

x	y
21.5	4.32
32.6	4.203
43.7	4.090
54.8	3.980

x	y
−3	1.1
−2.5	1.65
−2	2.475
−1.5	3.7125

x	y
1	0.8
3	7.2
6	194.4
8	1749.6

4. Unfortunately, the above property does not hold for functions of the form $y = Ab^x + C$. What goes wrong? For what kinds of values of A, b, and C does the property come close to being true?

GROUP WORK 2, SECTION 8.1

Comparisons

You have learned that an exponential function grows faster than a polynomial function. Find the values of $x > 0$ for which

1. $2^x \geq x^3$.

2. $(1.1)^x \geq x^2$.

3. $2^x \geq x^5$.

4. $3^x \geq x^5$.

5. $2^x > x^{10}$.

8.2 THE NATURAL EXPONENTIAL FUNCTION

▼ Suggested Time and Emphasis

$\frac{1}{2}$–1 class. Essential material.

▼ Points to Stress

1. The base e.

2. Continuously compounded interest.

▼ Sample Questions

- **Text Question:** Sketch a graph of $y = e^x$.

  Answer:

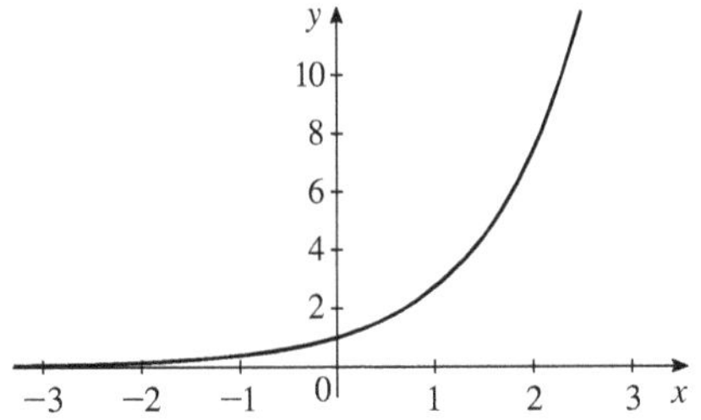

- **Drill Question:** Find the total amount of money in an account after 2 years if \$100 is invested at an interest rate of 5.5% per year, compounded continuously.

  Answer: $\$111.63$

▼ In-Class Materials

- Have your students fill out the following table, using their calculators, to give them a feel for $y = e^x$.

x	2^x	3^x	e^x
-2	$\frac{1}{4} = 0.25$	$\frac{1}{9} \approx 0.111$	≈ 0.135
-1	$\frac{1}{2} = 0.5$	$\frac{1}{3} \approx 0.333$	≈ 0.368
0	1	1	1
1	2	3	≈ 2.718
2	4	9	≈ 7.389
3	8	27	≈ 20.086

- Anticipate the next section by having students sketch the graphs of the inverse functions of 2^x and e^x by reflecting them across the line $y = x$.

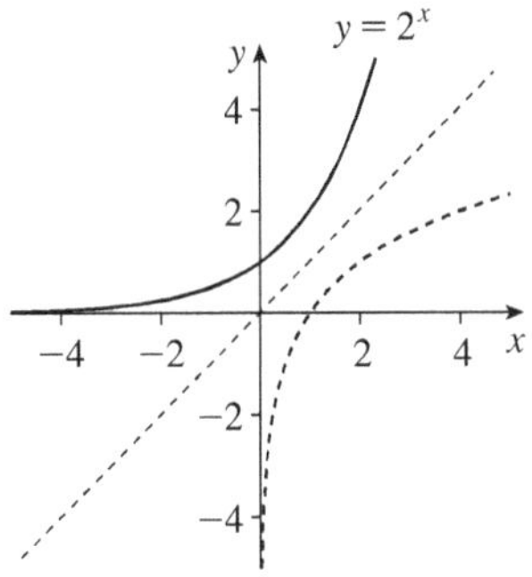

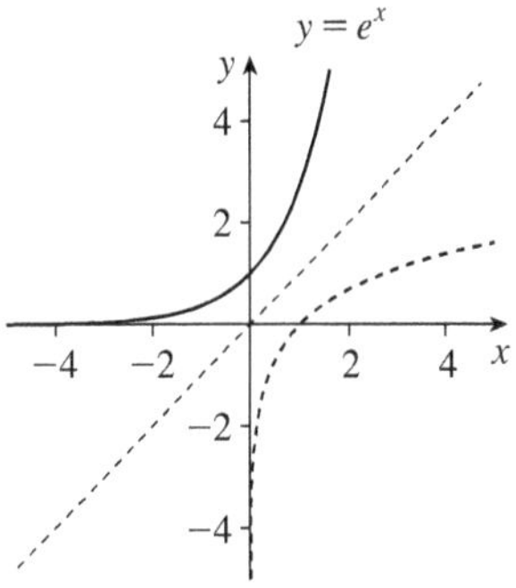

▼ Examples

- A shifted exponential curve:

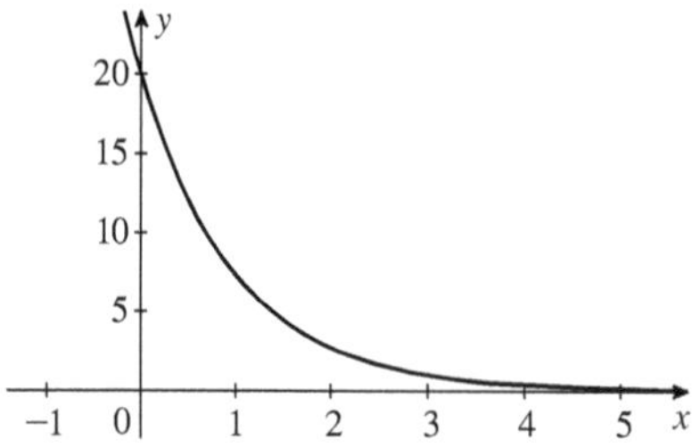

$f(x) = e^{-x+3}$

- A continuation of the previous section's analysis: \$2000 is put in an IRA that earns 7%. Its worth after 10 years is given in the following table.

Compounded annually	\$3934.30
Compounded monthly	\$4019.32
Compounded daily	\$4027.24
Compounded hourly	\$4027.50
Compounded continuously	\$4027.51

Note: Most calculators will not be able to determine the amount earned if the interest is compounded, say, every tenth of a second. Underflow errors in the microprocessor will make the computation difficult.

▼ Group Work 1: Finding a

The way we approximate e in this exercise foreshadows the computation that comes up when the derivative of e^x is used in calculus. Hand out the first page. Ideally, the students will realize that the value of a they are looking for in part 3 is between 2 and 3. If they don't, they may require the hint that part 1 implies that 2 is too small, and part 2 implies that 3 is too big. (These facts are not proved in the activity.)

After the students get close to 2.7, hand out the second page.

Answers:

1. (a) No, because $h = 0$ results in a zero denominator. **(b)** 1, 0.7177, 0.6956, 0.6934 **(c)** 0.6931

2. (a) No, because $h = 0$ results in a zero denominator. **(b)** 2, 1.1612, 1.1047, 1.0992 **(c)** 1.0986

3. 2.7 **4.** 0.9999 **5.** 1

▼ Homework Problems

Core Exercises: 2, 4, 8, 11, 17, 22, 31

Sample Assignment: 2, 4, 5, 8, 11, 14, 16, 17, 22, 23, 26, 31, 37

GROUP WORK 1, SECTION 8.2

Finding a

There is a problem that looks kind of silly on the face of it, but turns out to be very important in the development of calculus. In this activity we introduce the problem, and attempt to find a solution. Please present all answers to at least four decimal places unless directed otherwise.

1. Consider the function $f(h) = \dfrac{2^h - 1}{h}$.

(a) Is $f(0)$ defined? Why or why not?

(b) Compute $f(1)$, $f(0.1)$, $f(0.01)$, and $f(0.001)$.

(c) As h gets closer and closer to zero, does f approach a value? If so, what is that value?

2. Repeat part 1 using $f(h) = \dfrac{3^h - 1}{h}$.

3. It turns out to be important to find a number a such that as h gets closer and closer to zero, $f(h) = \dfrac{a^h - 1}{h}$ gets closer and closer to 1. Try to find such a value of a, accurate to one decimal place.

4. Now let $a = 2.718$. As h gets closer and closer to zero, what does $f\,(h) = \dfrac{2.718^h - 1}{h}$ approach?

5. Using the e^x key on your calculator, repeat part 4 for $f\,(h) = \dfrac{e^h - 1}{h}$.

8.3 LOGARITHMIC FUNCTIONS

▼ Suggested Time and Emphasis

1 class. Essential material.

▼ Points to Stress

1. Definition of the logarithm function as the inverse of the exponential function, from both a numeric and geometric perspective.
2. Properties of the logarithm function, emphasizing the natural and common logarithms.

▼ Sample Questions

- **Text Question:** It is a fact that $10^{\pi} \approx 1385.46$. Is it possible to approximate $\log 1385.46$ without the use of a calculator? If so, then approximate this number. If not, why not?
  Answer: $\log_{10} 1385.46 \approx \pi$
- **Drill Question:** Compute $\log_4 \frac{1}{64}$
  Answer: -3

▼ In-Class Materials

- When the logarithm function is graphed on a calculator, it appears to have a horizontal asymptote. Point out that the graph is misleading in that way. Start a graph of $y = \log_{10} x$ on the blackboard, noting the domain and the vertical asymptote. Using the scale of 1 inch = 1 unit (on the x-axis) and 1 foot = 1 unit (on the y-axis), plot some points:

x	0.1	1	2	3	4	5	6	7	8	9	10
$\log_{10} x$	-1	0	0.30	0.47	0.60	0.70	0.78	0.85	0.90	0.95	1

  Now ask how many inches we would have to go out to get up to $y = 2$ feet. (Answer: 100 inches, or $8\frac{1}{3}$ feet.) If the blackboard is large enough, plot the point $(100, 2)$. Then ask how far we would have to go to get up to $y = 5$ feet. (Answer: 1.57 miles.) Note that it takes close to a mile and a half to go from $y = 4$ to $y = 5$, and that (if you graphed it out) it would look a lot like there is a horizontal asymptote. Find the distance from your classroom to a city or landmark in another state, and ask the class to estimate the log of that distance, using the same scale.

- Sketch a graph of $f(x) = 2\log_2(x + 3)$. Sketch the inverse function, then find an algebraic formula for the inverse. Foreshadow the next section by showing that the graph of $f(x)$ is the same as that of $g(x) = \log_2\left((x + 3)^2\right)$.

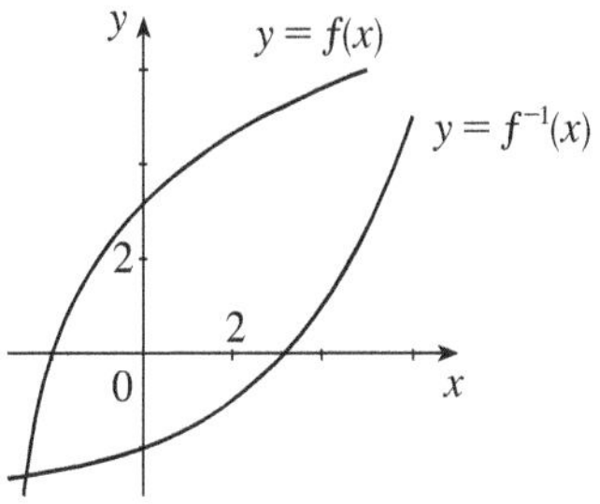

$f^{-1}(x) = 2^{x/2} - 3$

- Ask your students if they have ever had to deal with recharging the battery for a music device or a laptop. Assume that the battery is dead, and it takes an hour to charge it up half way. Ask them how much it will be charged in two hours. (The answer is 75% charged.) They may have noticed that, when charging a laptop battery, it takes a surprisingly long time for the monitor to change from "99% charged" to "100% charged." It turns out that the time it takes to charge the battery to $n\%$ is given by $t = -k \ln\left(1 - \frac{n}{100}\right)$; in our example $k = 1.4427$. Have students compute how long it would take the battery to get a 97% charge, a 98% charge, and a 99% charge. (Remind them that it took only an hour to go from 0% to 50%.) Graph t versus n to demonstrate that the battery will never be fully charged according to this model.

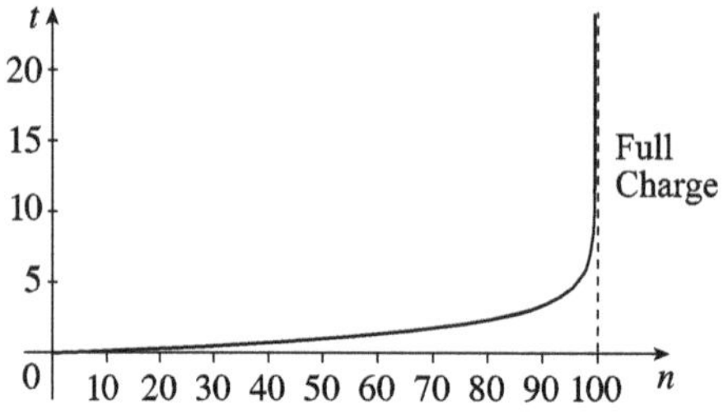

- Show students semilog graph paper (available at university bookstores, from your friendly neighborhood physics teacher, or from websites such as `http://www.csun.edu/science/ref/measurement/data/graph_paper.html`.) Point out how the distance between the y-axis lines is based on the logarithm of the y-coordinate, not on the y-coordinate itself. Have them graph $y = 2^x$ on semilog graph paper.

▼ Examples

A shifted logarithmic curve:

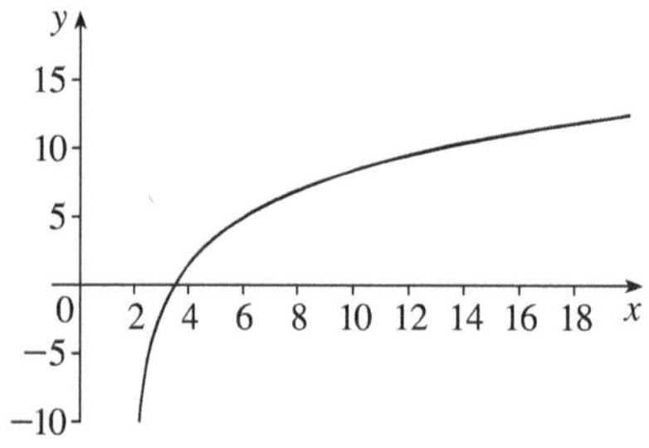

$y = 5 \ln(x - 2) - 2$

▼ Group Work 1: The Evan Operation

This activity can be enjoyed on two levels. One level of thought would stop after students used trial and error to discover that the Evan Function concatenates any two numbers. A more challenging level of thought would get students to verbalize how the Evan function works: why it does what it does. An advanced class might be given the goal, and challenged to come up with an expression for the Evan function themselves.

Problem 4 can serve as a springboard for discussions of commutativity, associativity, and identity, or it can stand on its own.

In this exercise, $[\![x]\!]$ denotes the greatest integer function.

Answers:

1. (a) 15 **(b)** 42,134 **(c)** 5,231,553 **(d)** 8,501,245 **(e)** 4,226,483,173

2. Evaning is the same as concatenating (writing one next to the other).

3. It is not possible, because negative numbers are outside the domain of the log function

4. (b) and (d) are true, (a) is false, and (c) is not even well defined.

5. $[\![\log_{10} b]\!] + 1$ tells us how many digits are in b. We then take a, add that many zeros to its end, and then add b.

▼ Group Work 2: Learning Curves

Phrases like "the learning curve is steep" and "the learning curve is shallow" come up occasionally in academia, and many don't know what they mean. This activity will use the idea of a learning curve to play with some of the functions that have been learned thus far. Either start the activity by handing out the worksheet, or by doing the first problem as a class, and then handing out the sheet.

Answers:

1.

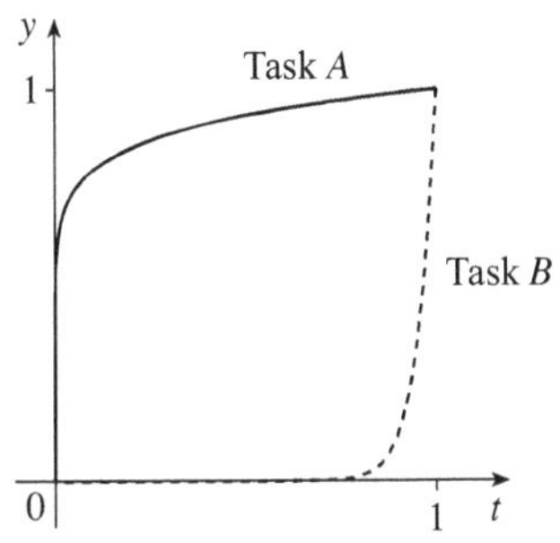

2. Task A. After saying it a couple of times, you pretty much know it.

3. Task B. There is a long period where improvment is slow, and then the learning takes place quickly.

4., 5. Answers will vary. Notice that for most activities of this nature, even though there is no horizontal asymptote, the slope of the learning curve will tend to approach zero (as happens for $y = \ln x$).

6. (a) Make sure students address the horizontal asymptote at $y = 1$ and the concavity change.

(c) Make sure students address the fact that the y-intercept is not zero.

(d) Make sure that students address the fact that there is no horizontal asymptote, and that improvement gets slower and slower (perhaps as discussed in Problem 5).

(f) Make sure students address the periodic "plateaus" in improvement.

(g) Make sure students address the "dips" where they get worse before they get better. (Games such as darts work like this, where as someone gets better, that person gets less and less lucky "fallout".)

7. It means that improvement is very rapid, implying that the task is easy to learn.

▼ Homework Problems

Core Exercises: 3, 6, 10, 13, 21, 32, 46, 65, 87

Sample Assignment: 3, 4, 6, 10, 11, 13, 17, 21, 27, 32, 36, 38, 46, 56, 65, 68, 87

GROUP WORK 1, SECTION 8.3

The Evan Operation

Consider two positive integers a and b. We know how to compute $a+b$, $a-b$, a/b, and ab. There is another thing we can compute: $a \circledast b$, pronounced "a Evan b". To Evan two numbers together, we perform the following calculation:

$$a \circledast b = a\left(10^{[\![\log_{10} b]\!]+1}\right) + b$$

1. Compute the following without the use of a calculator.

(a) $1 \circledast 5$

(b) $42 \circledast 134$

(c) $5231 \circledast 553$

(d) $8 \circledast 501{,}245$

(e) $42{,}264 \circledast 83{,}173$

2. Describe, in words, what it means to Evan two integers together.

3. What happens when we try to Evan negative integers?

4. Which of the following statements are true for arbitrary positive integers a, b, and c?

(a) $a \circledast b = b \circledast a$

(b) $a \circledast (b \circledast c) = (a \circledast b) \circledast c$

(c) $a \circledast 0 = a$

(d) $0 \circledast b = b$

5. Revisit your answer to Part 2. Explain, in words, why $a\left(10^{[\![\log_{10} b]\!]+1}\right) + b$ does what you say it does.

GROUP WORK 2, SECTION 8.3

Learning Curves

In psychology, we often graph the amount of material learned as a function of time spent studying or practicing. For example, assume I wanted to learn the capitals of all fifty states in the United States. I know the capitals of five states off the top of my head, and after ten minutes I can probably learn ten more. Ten more minutes will give me ten more, and then I'm going to slow down—I'll start forgetting. But if I put a solid three hours into it, I will know all fifty. If we plot these data on a graph, then I have a "learning curve".

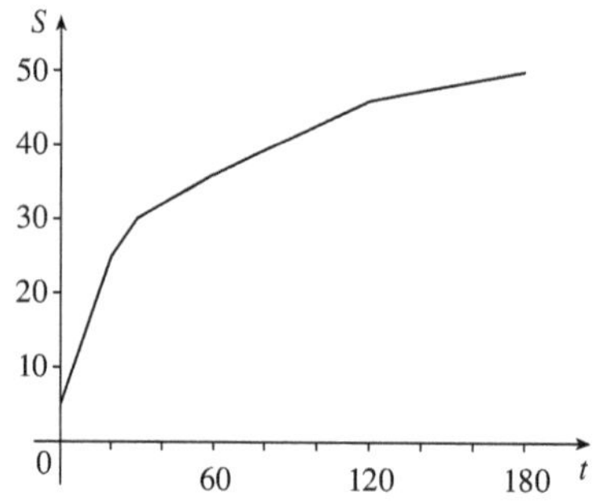

Different types of tasks have different learning curves, and, of course, the learning curves for each person are different. But for some tasks, the overall shape is similar for many people.

Let's look at two activities we are trying to learn how to do. Task A's curve is given by $y = \sqrt[10]{t}$ and Task B's is given by $y = t^{20}$.

1. Sketch both learning curves.

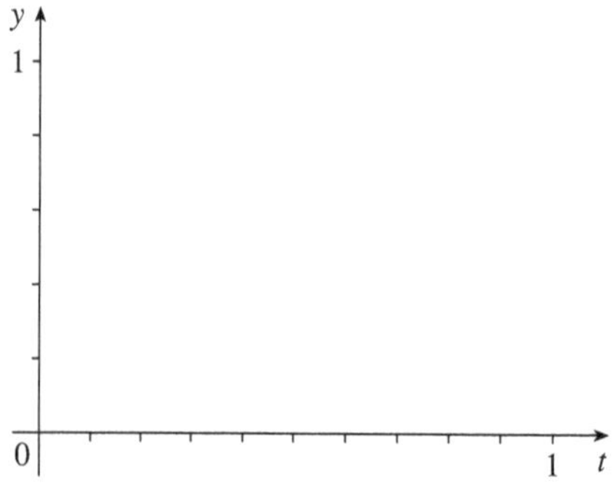

2. One of the tasks is memorizing a five digit number. Is this most likely Task A or Task B? Why?

3. It was interesting watching my five-year-old daughter learn to ride a bicycle. She experienced a period of frustration, where her improvement was slow. She couldn't balance. Suddenly, her mind and body seemed to "click" and she was able to ride perfectly, and started asking me how to pop wheelies, if she could ride to kindergarten by herself, and if unicycles were expensive. Did her learning curve look more like that of Task A or that of Task B? Why?

4. Both of the above tasks stopped at $y = 1$. We can think of $y = 1$ as "knowing the number" and "being able to ride down the street without falling off" Give an example of a task where there is no y-boundary; where one can start learning, and get better and better at the task without limit.

5. Draw the learning curve of the task you described in Part 4.

6. Come up with tasks that fit the following learning curves. Sketching their graphs, starting at $t = 0$, will be helpful.

(a) $L(t) = \dfrac{1}{1 + e^{-t+4}}$

(b) $L(t) = \dfrac{t}{10}$

(c) $L(t) = \dfrac{t}{10} + \dfrac{1}{2}$

(d) $L(t) = \ln(t + 1)$

(e) $L(t) = t^2$ for $0 \leq t \leq 1$

(f) $L(t) = x + \sin x$ for $0 \leq x \leq 9.5$

(g) $L(t) = x + 1.25 \sin x$ for $0 \leq x \leq 9$

7. If someone describes a task to you and says, "The learning curve is steep," what does that mean?

8.4 LAWS OF LOGARITHMS

▼ Suggested Time and Emphasis

$1\text{–}1\frac{1}{2}$ classes. Essential material.

▼ Point to Stress

The laws of logarithms, including the change of base formula.

▼ Sample Questions

- **Text Question:** Given that $\log_2 3 \approx 1.58496$, approximately what is $\log_2 9$?
  Answer: $\log_2 9 \approx 3.16992$

- **Drill Question:** Express $5\log(x+2) - \frac{1}{3}\log x$ as a single logarithm.
  Answer: $\log\left(\dfrac{(x+2)^5}{\sqrt[3]{x}}\right)$

▼ In-Class Materials

- Make sure your students do not neglect the warning after Example 4. Perhaps have them write out all the rules of logarithms they have learned so far, organized this way:

The Equivalence Definition	$y = \log_m x$ is equivalent to $x = m^y$
The Conversion Rule	$\log_m(\text{thing}) = \dfrac{\log_n(\text{thing})}{\log_n(m)}$
The Combining Rules	$\log_m\left(m^{\text{thing}}\right) = \text{thing}$ $m^{\log_m(\text{thing})} = \text{thing}$
The Arithmetic Rules	$\log_m 1 = 0$ $\log_m(ab) = \log_m a + \log_m b$ $\log_m\left(\frac{a}{b}\right) = \log_m a - \log_m b$ $\log_m\left(a^b\right) = b\log_m a$
The Non-Rules	$\log_m(a+b) = \log_m(a+b)$ $\dfrac{\log_m a}{\log_m b} = \dfrac{\log_m a}{\log_m b}$ $(\log_m a)^b = (\log_m a)^b$

  Perhaps if your students memorize the non-rules, they will be less likely to indulge in algebraic mischief under exam pressure.

- Note, without necessarily emphasizing, the importance of domains when applying the log rules. For example, $\ln\left(\frac{-5}{-6}\right)$ is not equal to $\ln(-5) - \ln(-6)$.

- Mention how logarithms were used to do calculations before the advent of the calculator. For example, every scientist had tables like these:

x	$\ln x$
4.4	1.48160
4.5	1.50408
4.6	1.52606
4.7	1.54756
4.8	1.56862

x	$\ln x$
9.8	2.28238
9.9	2.29253
10.0	2.30259
10.1	2.31254
10.2	2.32239

To find $\sqrt[3]{100}$, for example, the scientist would write

$$\begin{aligned} \sqrt[3]{100} &= 10^{2/3} \\ \ln\left(\sqrt[3]{100}\right) &= \ln\left(10^{2/3}\right) \\ &= \tfrac{2}{3}\ln(10) \end{aligned}$$

Then he or she would look up $\ln(10)$ from the table to get

$$\begin{aligned} \ln\left(\sqrt[3]{100}\right) &\approx \tfrac{2}{3}(2.30259) \\ &\approx 1.53506 \end{aligned}$$

And to find $\sqrt[3]{100}$, he or she would try to find 1.53506 in the right-hand column of the table. The result is that $\sqrt[3]{100}$ is between 4.6 and 4.7. In practice, the distance between table entries was much closer than 0.1. For a quicker, less accurate estimate, a slide rule was used.

▼ Examples

- A set of logarithms that can be combined:

$$a\ln(b+c) + \frac{3}{4}(\ln a - \ln b) = \ln\left((b+c)^a \sqrt[4]{\left(\frac{a}{b}\right)^3}\right)$$

- A set of logarithms that can be expanded:
  Given $\ln 2 \approx 0.693$ and $\ln 3 \approx 1.099$, compute $\ln \frac{8}{9}\sqrt[5]{6}$.

$$\begin{aligned} \frac{8}{9}\sqrt[5]{6} &= \frac{2^3}{3^2}(2\cdot 3)^{1/5} \\ \ln\left(\frac{8}{9}\sqrt[5]{6}\right) &= 3\ln 2 - 2\ln 3 + \tfrac{1}{5}(\ln 2 + \ln 3) \\ &\approx 0.240569 \end{aligned}$$

▼ Group Work 1: Irrational, Impossible Relations

One of the joys of precalculus is that there are some deep, graduate-level mathematical results that can be explored using nothing more than the techniques of algebra and one's coconut. Start by asking students this true-or-false question:

$$\log_2 3 = \frac{4953}{3125}$$

The answer is, of course, "false". Next, review the definitions of rational and irrational numbers, and hand out the first sheet. The hint sheet should be given out only after students have tried to show that $\log_2 3$ is irrational, or at least discussed it enough to understand what they are trying to show. If a group finishes early, have them show that $\log_2 a$ is always irrational if a is an odd integer.

Answers:

1. $-\gamma$ **2.** $-\gamma$ **3.** $\frac{\gamma}{2}$ **4.** $\log_2 3 = \frac{\ln 3}{\ln 2}$

5. The proof is outlined in the hint sheet.

Answers (Hint Sheet):

1. $2^{a/b} = 3 \quad \Rightarrow \quad \sqrt[b]{2^a} = 3 \quad \Rightarrow \quad 2^a = 3^b$

2. If $a = b = 0$, then $\log_2 3 = 0/0$, which is undefined.

3. $2^a = 2 \cdot 2 \cdot \cdots \cdot 2$, and $3^b = 3 \cdot 3 \cdot \cdots \cdot 3$, so these numbers can never be equal, because the left is always divisible by two, and the right never is (unless $b = 0$).

4. It is irrational because it cannot be written as a ratio of two integers.

▼ Group Work 2: The Slide Rule

If possible, give each group a slide rule. If you don't happen to have a lot of real slide rules around, a slide rule cutout is provided. Teach the students how to use the slide rule to multiply:

To multiply a times b, move the 1 on the top over the a on the bottom. Then locate the b on the top, and right below it will be ab. Below is an example, where we multiply 6 by 8 to get 48.

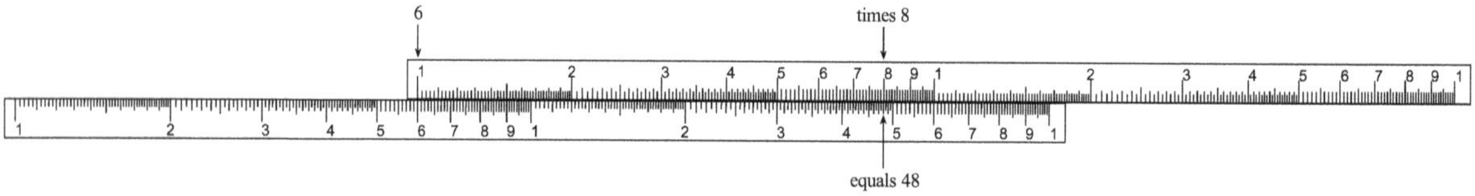

Accuracy isn't the main issue here (real slide rules have a very thin guide that slides perpendicular to the rules, with an embedded hair to ensure accuracy.) The idea is to see how it works.

After students have learned to multiply, hand out the activity sheet.

Answers:

Problems 1–5 are trivial. Note that Problem 5 can be done by multiplying 8×1.2 and remembering to multiply the answer by 10.

6. The numbers are not equally spaced. In fact, they are spaced based on the log of the number. In other words, the 2 is log 2 units from the 1, the 3 is log 3 units from the 1, the 4 is log 4 units from the 1, etc. That is why the numbers get closer and closer together. When you slide the top, you are in effect *adding* the two logarithms. So when we move the slide over the six, and look at the eight, we are adding $\log 6 + \log 8$ The total distance is then $\log(6 \times 8) = \log 48$. And when we look below the 8, we see the number that is log 48 units from the 1, which is 48. In short: We are adding logs, which in effect multiplies the numbers.

▼ Homework Problems

Core Exercises: 1, 2, 9, 14, 28, 33, 46, 48, 55

Sample Assignment: 1, 2, 3, 9, 14, 18, 28, 33, 35, 44, 46, 48, 53, 55, 72

GROUP WORK 1, SECTION 8.4

Irrational, Impossible Relations

1. If $\log_2 x = \gamma$, then what is $\log_{1/2} x$?

2. If $\log_b x = \gamma$, then what is $\log_{1/b} x$ (assuming $b > 1$)?

3. If $\log_b x = \gamma$, then what is $\log_{b^2} x$?

4. We are going to estimate $\log_2 3$. Of course, in fifth grade, you memorized that $\log_2 3 \approx 1.584962501$. Suppose you didn't have this fact memorized. There is no $\log_2 3$ button on your calculator! How would you compute it?

5. Unfortunately, the calculator gives us only a finite number of digits. If $\log_2 3$ were a rational number, we would be able to express it as a fraction, giving us perfect accuracy. Do you think it is rational or irrational? Try to prove your result.

GROUP WORK 1, SECTION 8.4

Irrational, Impossible Relations (Hint Sheet)

So, you realize that it's not easy to determine whether $\log_2 3$ is rational!

One way to attempt to show that $\log_2 3$ is rational is to assume that it is, and try to find integers a and b such that $\log_2 3 = \frac{a}{b}$. If we can show that there are no such a and b, then $\log_2 3$ *cannot* be rational.

1. Assume that $\log_2 3 = \frac{a}{b}$. Show that a and b must then satisfy $2^a = 3^b$

2. Notice that $a = 0$, $b = 0$ satisfies $2^a = 3^b$. Show that this fact doesn't help us.

3. Find $a \neq 0$ and $b \neq 0$ that satisfy $2^a = 3^b$, or show that no such $\{a, b\}$ exists.

4. Is $\log_2 3$ rational or irrational? Why?

GROUP WORK 2, SECTION 8.4

The Slide Rule

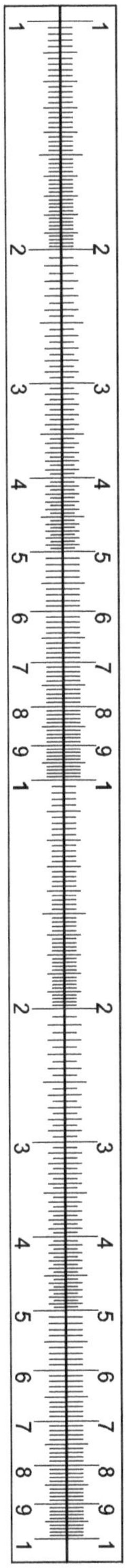

Before we ask the million-dollar question, it is important that you are familiar with the operation of a slide rule. Use your slide rule to do the following computations. (We are aware that you don't *need* a slide rule to do them, the idea here is to learn how a slide rule works.)

1. $2 \times 4 =$

2. $5 \times 5 =$

3. $1.3 \times 8 =$

4. $6 \times 8 =$

5. $8 \times 12 =$

And now, the big question:

6. Why does the slide rule work? What is this magical property it possesses that allows us to multiply in this way?

Hint: It has something to do with the section you have just covered in class.

8.5 EXPONENTIAL AND LOGARITHMIC EQUATIONS

▼ Suggested Time and Emphasis

1 class. Essential material. Can be combined with Sections 8.6 and 8.7.

▼ Point to Stress

Solving equations involving exponential and logarithmic functions, algebraically and graphically.

▼ Sample Questions

- **Text Question:** Solve the equation $4 + 3\log(2x) = 16$.
  Answer: $x = 5000$
- **Drill Question:** If I invest \$2000 at an annual interest rate of 3%, compounded continuously, how long will it take the investment to double?
  Answer: $\frac{\ln 2}{0.03} \approx 23$ years

▼ In-Class Materials

- At this point students know the algebraic rules for working with exponential and logarithmic functions. Stress that if these rules do not suffice to solve an equation, there is a good chance they cannot find an exact solution. Give students an equation such as $x^2 = 2\ln(x+2)$ (Example 9) and have them try to solve it algebraically. The correct answer is that students cannot do so, but you will find that many make up rules and somehow wind up with a solution.
- Review the concept of inverse functions. Have students find the inverse of functions such as $f(x) = 2^{x^3+1}$ and $f(x) = \ln(x-5) + e^3$.
- Build up some tough problems from simple ones. For example, first have students solve $x^2 + 2x - 15 = 0$. Then have them solve $(e^x - 1)^2 + 2(e^x - 1) - 15 = 0$. One can go even further with $\ln\left((e^x - 1)^2 + 2(e^x - 1) - 14\right) = 0$.

▼ Examples

- Solve $\log(x^2 - 1) - \log(x + 1) = 3$.
  Answer: $x = 1001$
- Solve $(\ln x - 2)^3 - 4(\ln x - 2) = 0$.
  Answer: $x = 1, e^2, e^4$

▼ Group Work 1: The Rule of 72

If a group finishes early or has trouble with Problem 7, have them redo Problems 1–5 for annual compounding. Another good continuation is the following: is it true that the length of time it will take an investment to quadruple is always twice the amount that it will take the money to double? Prove your answer or provide a counterexample. The proofs that the students write may surprise you with their quality and use of the properties of exponents and logarithms.

Answers:

1. 13.86 years
2. The doubling time is not affected by changes in principal. Algebraically, the P_0 drops out of the exponential growth equation. Intuitively, the doubling time is a property of the ratio of two numbers, not a property of the numbers themselves.
3. Estimate: 14.4 years, a 3.90% error.
4.

Interest Rate	Actual Doubling Time	Estimated Doubling Time	Error
3%	23.1 years	24 years	3.90%
8%	8.66 years	9 years	3.92%
12%	5.78 years	6 years	3.81%
18%	3.85 years	4 years	3.90%

The Rule of 72 is accurate to within 4% for the given range of interest rates.

5. If r is the interest rate (as a number, not a percentage), then the doubling time is $D = \dfrac{\ln 2}{r}$.
6. $100 \ln 2 \approx 69$. This estimate gives a doubling time of 13.8, for an error of 0.4%.
7. If the compounding is assumed to be annual, 72 works best for values of r between 3% and 15%. For example, 10% compounded annually doubles every 7.2735 years. The rule of 72 gives 7.2, a good approximation, and the rule of 69 gives 6.9, a bad one.

▼ Group Work 2: Every Nest Egg Needs a Bird

This activity involves some repetitive calculation. Stress that part of working in a group effectively is to figure out how best to divide up the work.

Answers:

1. $2000e^{(0.07)47} + 2000e^{(0.07)46} + 2000e^{(0.07)45} + 2000e^{(0.07)44} + 2000e^{(0.07)43} + 2000e^{(0.07)42}$
$+ 2000e^{(0.07)41} + 2000e^{(0.07)40} + 2000e^{(0.07)39} + 2000e^{(0.07)38} = \$399{,}759.00$
2. $\sum_{n=0}^{37} 2000e^{(0.07)n} = \$366{,}753.00$
3. Most people are surprised by this fact.

▼ Homework Problems

Core Exercises: 3, 10, 23, 30, 38, 45, 60, 73, 76

Sample Assignment: 3, 10, 23, 26, 30, 35, 38, 45, 51, 55, 60, 70, 73, 76, 81, 84

GROUP WORK 1, SECTION 8.5

The Rule of 72

In this exercise, we attempt to answer the question asked by many investors: "How long is it going to take for me to double my money?"

1. Consider an investment of \$100 invested at 5%, compounded continuously. How long would it take for the investor to have \$200?

2. What would the doubling time be if the initial investment were \$1,000? \$10,000? What effect does changing the principal have on the doubling time, and why?

One of the first things that is taught in an economics class is the Rule of 72. It can be summarized thusly:

"The number of years it takes an investment to double
is equal to 72 divided by the annual percentage interest rate."

3. What would the Rule of 72 say the doubling time of a 5% investment is? Is it a good estimate?

4. Repeat Problems 1 and 3 for investments of 3%, 8%, 12% and 18%. What can you say about the accuracy of the Rule of 72?

5. Derive a precise formula for the time T to double an initial investment.

6. There is an integer that gives a more accurate answer for continuous or nearly continuous compounding than the Rule of 72. What is this number? Check your answer by using it to estimate the doubling time of a 5% investment.

7. It turns out that there is a reason that we use the number 72 in the rule. It has to do with one of the assumptions we made. Why do economists use the Rule of 72?

GROUP WORK 2, SECTION 8.5

Every Nest Egg Needs a Bird

Johnny and Jenny had a crazy dream. They wanted to retire wealthy, or at least comfortable.

1. Johnny started saving at age 20. Every year, for simplicity assume January 1 of every year, Johnny put \$2000 into a type of personal retirement savings account called an IRA. Again, for simplicity, let's assume that the IRA made 7% interest per year. Ten years later, Johnny gave up saving for retirement, and never put any money away again. How much money did Johnny have when he was 67 years old?

2. Jenny spent most of her twenties reading magazines and watching movies, and didn't start saving until she was 30. At that point she started putting \$2000 a year into her IRA (again, let's assume it makes 7% per year and she put it away on January 1). When she was 40 she continued to save, and at 50 she continued to save. In fact, she put in \$2000 per year until she was 67, when she also retired. How much money did Jenny have when she was 67?

3. Which of the two had more money? Is that the result you expected?

8.6 MODELING WITH EXPONENTIAL AND LOGARITHMIC FUNCTIONS

▼ Suggested Time and Emphasis

$\frac{1}{2}$–1 class. Recommended material. Can be combined with 8.5.

▼ Points to Stress

1. Translating verbal descriptions of problems into mathematical models, and solving the problems using the models.

2. Certain standard types of problems such as those dealing with exponential growth and decay and logarithmic scales.

▼ Sample Questions

- **Text Question:** Recall that Newton's Law of Cooling is given by

$$T(t) = T_s + D_0 e^{-kt}$$

Which of the constants in this law correspond to surrounding temperature? Which represents the initial difference between the object and its surroundings? How do you know?
Answer: T_s, D_0. There is a horizontal asymptote at T_s, which would have to correspond to the surrounding temperature, because things cool off to the surrounding temperature. Their initial difference is D_0, because when $t = 0$ we know that $T = T_s + D_0$.

- **Drill Question:** Recall that the pH of a substance is given by $-\log[H^+]$, where H^+ is the concentration of hydrogen ions measured in moles per liter. Also recall that solutions with a pH of 7 are neutral, those with pH > 7 are basic, and those with pH < 7 are acidic.

  (a) If a sample of rebulon were measured to have hydrogen concentration of $[H^+] = 4 \times 10^{-8}$ M, what would the pH be?

  (b) What is the hydrogen ion concentration in a neutral substance?

  Answer: (a) 7.4 **(b)** 10^{-7}

▼ In-Class Materials

- Show that the expression $y = e^{kt}$ can be written as $y = a^t$ and vice versa. Add that e^{kt+c} is equivalent to Ae^{kt}.

- One way to measure the growth of the Internet is to measure the number of Internet hosts over time. Twenty years' worth of these data are shown below. Try to determine with students whether this is an example of exponential growth. (Note: Do not show the student the third column right away. Let them come up with the idea of finding growth rates between data points.)

Month	Hosts	Growth
Aug 1981	213	–
May 1982	235	1.1404
Aug 1983	562	2.0065
Oct 1984	1024	1.6701
Oct 1985	1961	1.9150
Feb 1986	2308	1.6217
Nov 1986	5089	2.8782
Dec 1987	28,174	4.8615
Jul 1988	33,000	1.3112
Oct 1988	56,000	8.1509
Jan 1989	80,000	4.1168
Jul 1989	130,000	2.6620
Oct 1989	159,000	2.2231
Oct 1990	313,000	1.9686
Jan 1991	376,000	2.0824
Jul 1991	535,000	2.0364
Oct 1991	617,000	1.7608
Jan 1992	727,000	1.9172
Apr 1992	890,000	2.2511
Jul 1992	992,000	1.5453
Oct 1992	1,136,000	1.7122

Month	Hosts	Growth
Jan 1993	1,313,000	1.7762
Apr 1993	1,486,000	1.6520
Jul 1993	1,776,000	2.0443
Oct 1993	2,056,000	1.7875
Jan 1994	2,217,000	1.3487
Jul 1994	3,212,000	2.1120
Oct 1994	3,864,000	2.0818
Jan 1995	4,852,000	2.4618
Jul 1995	6,642,000	1.8739
Jan 1996	9,472,000	2.0357
Jul 1996	12,881,000	1.8525
Jan 1997	16,146,000	1.5654
Jul 1997	19,540,000	1.4692
Jan 1998	29,670,000	2.2900
Jul 1998	36,739,000	1.5387
Jan 1999	43,230,000	1.3809
Jul 1999	56,218,000	1.6985
Jan 2000	72,398,092	1.6516
Jul 2000	93,047,785	1.6541
Jan 2001	109,574,429	1.3831

Answer: We can graph the data, and get a curve that looks like exponential growth. We can also graph growth rate and see (except for two spikes in the late 1980s) a more-or-less constant growth rate.

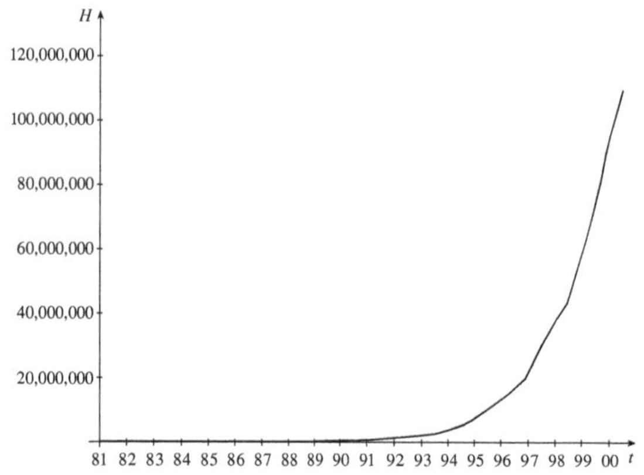

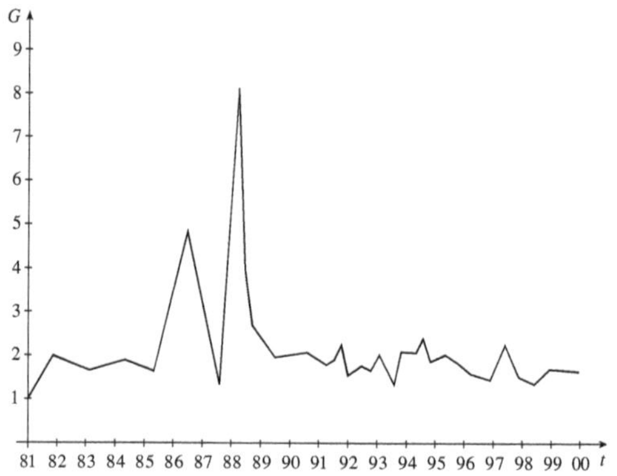

- In 1985 there were 15,948 diagnosed cases of AIDS in the United States. In 1990 there were 156,024 cases. Scientists said that if there was no research done, the disease would grow exponentially. Compute the number of cases this model predicts for the year 2000. The actual number was 774,467. Discuss possible flaws in the model with students, and point out the dangers of extrapolation.
- Discuss the logistic growth model $P = \dfrac{M}{1 + Ae^{-kt}}$. Have students graph a few of these curves with different values of M and k. This model assumes that an environment has a carrying capacity M. It assumes that when a population is much less than M, a population's growth will look like exponential growth, but that when the population approaches M, the population growth gets very slow, asymptotically approaching M. If the population starts out greater than M, then it will decay, exponentially, to M.
- Go over Examples 9 and 10, the Richter scale. Ask students the open-ended question, "How much worse is an earthquake that measures 7 on the Richter scale than an earthquake that measures 6?"

▼ Examples

Exponential decay: we know that the half-life of carbon-14 is 5,730 years. In 1988, the Vatican consented to give a few fibers to scientists to carbon date the Shroud of Turin, an ancient artifact. They found that the fibers had 92% of the original ^{14}C left. Discuss what this implies about the age of the shroud.

▼ Group Work 1: The Coffee Window

In physics and chemistry, people often refer to "Newton's Law of Cooling" and "Newton's Law of Heating". After completing this exercise, students should have discovered that these are the same law, the only difference being the relative temperatures of the substance being studied and its surroundings. The proper names have been taken from *The Hitchhiker's Guide to the Galaxy* by Douglas Adams. The temperatures were determined by painful experimentation.

This activity can also launch a discussion of how mathematical models are simplifications of reality. For example, in Newton's Law of Cooling, we assume that the temperature of the cup's surroundings is constant. But of course, that isn't exactly true. In fact, one could argue that as the coffee cools, the energy from the coffee heats up the surroundings by a fraction, and that isn't taken into account in our model. No mathematical model will ever take into account every effect on every molecule of the universe. The question to ask isn't whether a model is true or false, but whether it is accurate enough to be useful.

Answers:

1. The coffee cools down over time; it does not heat up.

2. The thermos is better. A smaller value of k corresponds to a smaller rate of change of the coffee's temperature.

3. $T = \left(118e^{-0.03t} + 42\right)^{\circ}$

4. $t \approx 6.19$ minutes

5. $t \approx 20.92$ minutes

6. $t \approx 14.73$ minutes

7. $t \approx 34.82$ minutes

8. It will never cool down to drinking temperature.

9.

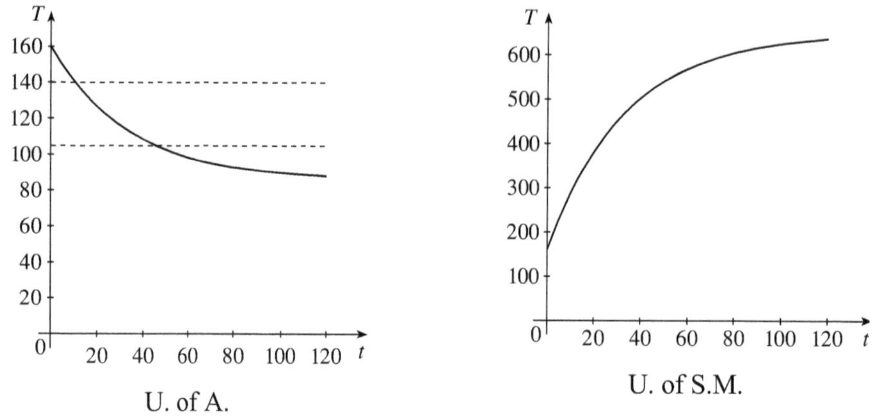

They would probably call it "Zaglor's Law of Heating".

▼ Group Work 2: Earthquake

This activity is based on Exercises 36 and 37 from the text.

Answers:

1. Approximately 2500 times as intense
2. Approximately twice as intense
3. When the magnitude of an earthquake increases by 0.3 on the Richter scale, its intensity doubles.

▼ Homework Problems

Core Exercises: 1, 4, 11, 16, 19, 26, 36, 40

Sample Assignment: 1, 4, 7, 11, 14, 16, 19, 21, 24, 26, 28, 31, 36, 40, 43

GROUP WORK 1, SECTION 8.6

The Coffee Window

(Written with John Hall)

Dr. Tricia MacMillan has a problem. Every day she leaves her apartment in London at the crack of dawn and heads for Milliway's, where she purchases a delicious cup of piping hot coffee. She drinks this coffee while walking to her office. The problem is that sometimes she burns her tongue badly with her first sip, while other times she waits too long and her coffee gets cold. The latter case is the worst, because besides doing a pretty bad job of keeping you warm, cold coffee tastes terrible. As it drops below a certain temperature, coffee undergoes a chemical reaction which turns even the most expensive brand into something that tastes absolutely filthy.

Being a mathematician, Dr. MacMillan doesn't just get mad, she gets more coffee and does an experiment. She wants to figure out exactly when she can take her first sip without burning herself, and from that point, how much time she has before the coffee turns bad. Every one of her mornings for the next week is spent in Milliway's with an oven thermometer and a cup of fresh coffee.

After much painful experimentation, Dr. MacMillan determines that if the temperature of the coffee is above $140°$ F, it burns her tongue. If the temperature drops below $105°$, the coffee undergoes the reaction and becomes undrinkable (unless she's already burnt her tongue so badly in the first experiment that she can't taste a thing).

Just like every other substance in the universe, coffee obeys Newton's Law of Cooling. Its temperature as a function of time is given by

$$T(t) = T_s + D_0 e^{-kt}$$

Note that there are three parameters in this equation. One is the outside temperature, and one depends on the initial temperature of the coffee. For a typical Styrofoam cup, $k \approx 0.05$, if t is measured in minutes.

1. Why is this constant positive?

Dr. MacMillan scoffs at Styrofoam. She is the proud owner of a Sirius Cybernetics Corporation thermos (only 35% asbestos!) For this thermos the constant is $k = 0.03$.

2. Which does a better job of keeping the coffee warm, the Styrofoam cup or the thermos? How does knowing the value of k allow you to figure out the answer?

The next day, Dr. MacMillan leaves Milliway's with a thermos full of coffee at 160° F. It is 8:30 A.M., and the outside temperature is 42°.

3. Find T_s and D_0, and rewrite $T(t) = T_s + D_0 e^{-kt}$ with the appropriate constants for this situation. (Let the time t be measured in minutes, and let $t = 0$ stand for 8:30 A.M.)

4. How long must she wait before she is able to drink the coffee?

5. At what time will the coffee fall below 105° and become undrinkable?

6. How much time does Dr. MacMillan have to drink her coffee?

7. What would the answer to Problem 6 be if Dr. MacMillan were teaching at the University of Arizona, where the outside temperature is 86°?

8. What would the answer to Problem 6 be if she were teaching at the University of Southern Mercury (outside temperature 650° in winter)?

9. Draw a graph of coffee temperature versus time for the University of Arizona and the University of Southern Mercury. Why do they look different? What would students at the University of Southern Mercury call Newton's "Law of Cooling" (if they spoke English)?

GROUP WORK 2, SECTION 8.6

Earthquake

---

1. The devastating 1906 earthquake in San Francisco had a magnitude of 8.3 on the Richter scale. At the same time, in Japan, an earthquake with magnitude 4.9 caused only minor damage. How many times more intense was the San Francisco earthquake than the Japanese earthquake?

2. The Alaska earthquake of 1964 had a magnitude of 8.6 on the Richter scale. How many times more intense was this than the 1906 San Francisco earthquake?

3. Fill in the blank: When the magnitude of an Earthquake goes up by ____ on the Richter scale, its intensity doubles.

8.7 DAMPED HARMONIC MOTION

▼ Suggested Time and Emphasis

1 class. Optional material.

▼ Points to Stress

- Recognizing situations where damped harmonic motion occurs.
- Properties of the functions $y = ke^{-ct} \sin(\omega t)$ and $y = ke^{-ct} \cos(\omega t)$, where $c > 0$.
- Using the above functions to create models of damped harmonic motion.

▼ Sample Questions

- **Text Question:** Consider the following equation $y = 100e^{-3t} \sin(10t)$ for damped harmonic motion.
  (a) What is the value of the damping constant?
  (b) What effect would increasing the damping constant have on the graph of the function?
  Answer: (a) 3 (b) It would cause the oscillation to die down more quickly.
- **Drill Question:**
  (a) Find a function of the form $y = ke^{-ct} \cos(\omega t)$ that models damped harmonic motion with initial amplitude 10, damping constant 2, and frequency 3.
  (b) Sketch a graph of this function.
  Answer:
  (a) $y = 10e^{-2t} \cos(6\pi t)$
  (b)

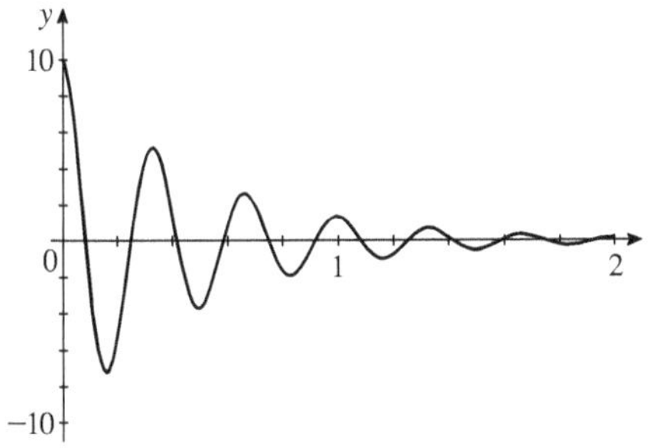

▼ In-Class Materials

- Go over some classic examples of harmonic motion and decide which, in the real world, would actually be best modeled by damped harmonic motion. Pay particular attention to any examples that were given when harmonic motion was covered. For example, a mass on a spring or swinging pendulum is damped harmonic motion, but the height of a point on a spinning wheel or alternating current is simple harmonic motion.
- The damping constant is often very small. As an example, bring in a tuning fork or a pendulum on a long string. Notice how the sound or swinging can take a long time to die out.
- Two of the damped oscillation constants are easy to determine experimentally. If you start a pendulum at the peak of its swing, you can determine k. The period can be determined with a watch, from which ω can be calculated. The constant c requires a little more work, but can be derived by measuring a second peak and noting the time. This kind of calculation will be practiced in the group work.

▼ Examples

- A baseline damped oscillation function, followed by functions obtained by varying each of the constants in turn.

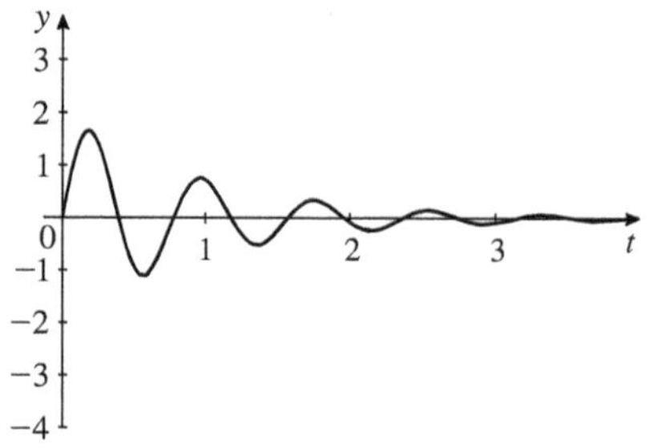

$y = 2e^{-t}\sin 8t$

$y = 4e^{-t}\sin 8t$

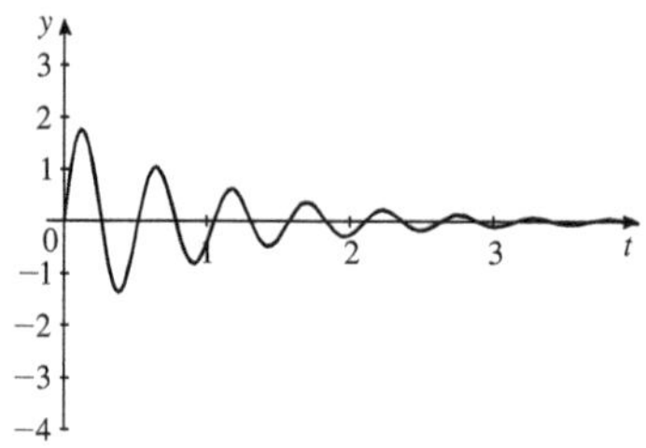

$y = 2e^{-t}\sin 12t$

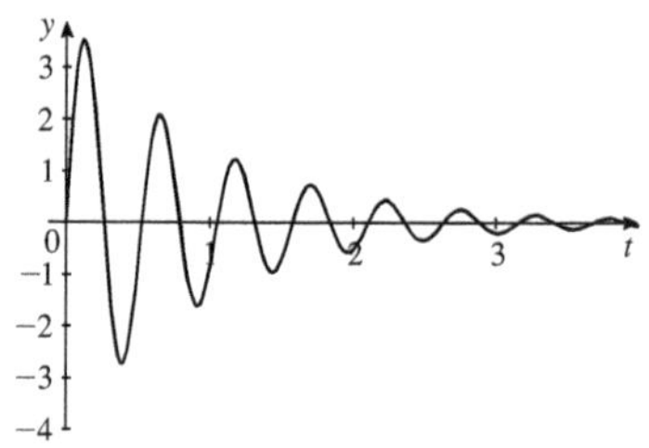

$y = 4e^{-t}\sin 12t$

▼ Group Work: Find the Constants

This activity is designed to give the students practice working with damped oscillation constants using a graphical approach.

Answers:

1. $y = ke^{-ct}\cos(\omega t)$
2. $k = 150$
3. $\omega = \frac{\pi}{2}$
4. $c = 0.55$ [Answers will vary, given that it is an estimate. Assuming the point $(2, -50)$ is on the curve gives $c \approx 0.5493061445$.]
5. Answers will vary. It would be harder to find the value of k because $(0, k)$ would not be on the graph.

▼ Homework Problems

Core Exercises: 4, 5, 8, 11, 17

Sample Assignment: 1, 4, 5, 7, 8, 11, 13, 16, 17

GROUP WORK, SECTION 8.7

Find the Constants

Consider the damped oscillation function shown below.

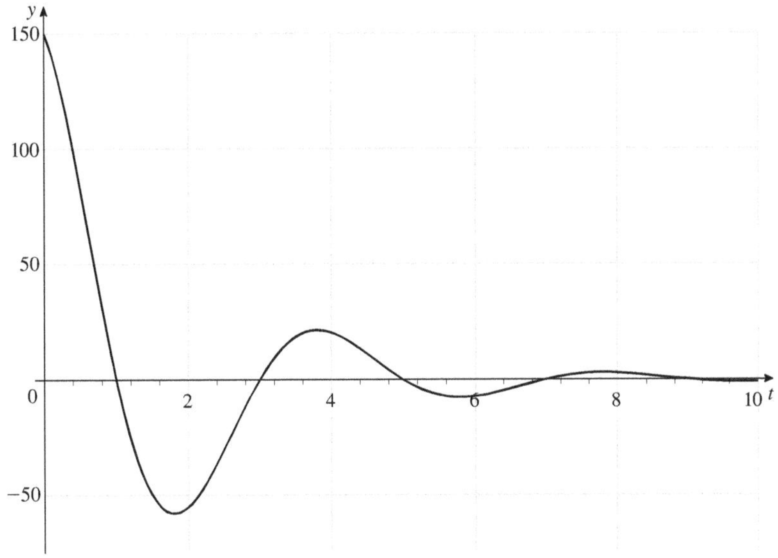

1. Is the equation of the form $y = ke^{-ct}\sin(\omega t)$ or $y = ke^{-ct}\cos(\omega t)$ (where $c > 0$)?

2. Compute the value of k.

3. Compute the value of ω.

4. Find a good estimate of the value of c.

5. Would this problem have been easier, harder, or the same for the graph shown below?

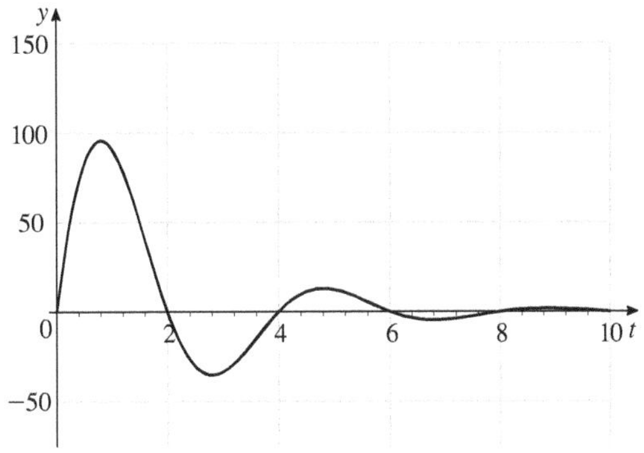

www.ingramcontent.com/pod-product-compliance
Ingram Content Group UK Ltd.
Pitfield, Milton Keynes, MK11 3LW, UK
UKHW050704091126
16428UKWH00045B/1427

9 781111 574765